THE COMPLETE AI GUIDE FOR SMALL & MEDIUM SIZED BUSINESSES

Navigating Change, Unlocking Value, and Leading with Purpose in the Age of Intelligent Machines

George Mathew

ISBN: 978-1-7643060-1-0

Cover design by: Mathew Stefan Thomas

Author Contact
George Mathew
email: contact@georgeebymathew.com
website: www.georgeebymathew.com
linkedin: https://www.linkedin.com/in/gem2006/

To Dips, the love of my life and Ash and Zac who complete me.

CONTENTS

Title Page

Copyright

Dedication

Preface

The AI Horizon 1

The Quiet Revolution 12

Busting the Myths 21

The AI Iceberg 30

The Art of Framing the Right Problem 40

Picking the Right AI Solution 53

Avoiding Pilot Purgatory 73

AI for SMBs 82

Leveraging What You Already Have 91

Bridging the Skills Gap 99

AI's Double-Edge 109

Guarding Your Data 116

Transformation Reimagined 125

Books By This Author 135
Praise For Author 137

PREFACE

If you're holding this book in your hands, it likely means one thing: you care about AI. You care about understanding how technology is shaping the world around us. You care about making choices that are not just smart, but meaningful. And you care about making a difference - whether for yourself, in your career, your organisation, or the broader world.

It was Father's Day, 7 September 2025, when I found myself on a Sunday afternoon panel discussion about AI. What made this event unusual was its backdrop: the Australian premiere of Doomers by Matthew Gasda - a play set in the boardroom of a successful AI company in 2023, dramatizing the firing of its CEO and the stark ethical and existential questions that followed. The theatre was packed with parents, artists, writers, students, and people from all walks of life - proof of how deeply AI has entered the public imagination.

I have never before witnessed such a convergence of public interest in a technology revolution and its simultaneous depiction in popular culture. Of course, Apple and Facebook have long since seeped into Hollywood and mainstream narratives, but this felt different. Not since the Y2K era of the late 1990s when, as a Gartner analyst, I knew the looming catastrophe would not materialise, have I seen such a charged mix of fear and anticipation. And yet, in the atmosphere of Tom Mann Theatre in Sydney, the sense of both hysteria and hope surrounding AI was palpable.

Since Y2K, I've spent over two decades navigating the frontlines of digital and more recently AI-led transformation across the world - from the bustling innovation hubs of North America and Europe to the fast-growing markets of Asia, the Middle East and the Pacific. I've seen both the dazzling potential of technology and the frustrations that come when it falls short of expectations. I've helped organisations unlock new revenue streams, streamline operations, and reimagine experiences, and I've witnessed my own AI projects fail - not because the technology wasn't capable, but because we didn't approach it thoughtfully. Yet at the time of writing I have had 3x more success than global averages in leading organisations past an AI pilot to full implementation in the markets I served.

This book is my attempt to bridge that gap between

promise and reality. It draws from my experience of having led over 50 business transformations of all sizes and shapes, triggered by many of the previous technology revolutions - ecommerce, cloud computing, analytics, mobile and social, collectively called digital, and more recently AI. It's not just a guide to AI or digital transformation. It's a conversation - one I wish I could have had early in my career - with anyone who wants to navigate the promise and complexity of this rapidly evolving AI world. My goal is simple: to help you make better decisions, avoid the costly missteps I've seen too many teams make, and those I have made myself and have learnt from to embrace the opportunities that truly matter. There is an amazing parallel that AI is taking as Digital did.

I have addressed small and medium sized business leaders particularly the C-suite and their direct reports. The larger corporations have deeper pockets and the resources to engage expensive consultants that you don't. You don't have to engage in long drawn transformations while running your business at the same time. So you'll find stories from the field - some of triumph, some of failure - alongside insights, frameworks, and perspectives that I've distilled over years of hands-on experience. But more than anything, I hope you'll feel the human side of innovation. Because at its core, AI, digital transformation, and even the most complex technologies are not just about machines - they are about people: our curiosity, our creativity, our ambition, and sometimes, our fear.

Recent studies show that while 90% of employees are already leveraging AI to boost their personal productivity, and an overwhelming 95% of generative AI investments fail to deliver measurable returns. For smaller businesses, however, extending AI safely from personal productivity into daily operations is both feasible and highly achievable. Unlike large organisations, which often struggle to move beyond pilots due to challenges such as fragmented data ownership, entrenched silos, and organisational resistance, smaller businesses can act with agility and focus. In many ways, they are better positioned to unlock tangible value from AI than their larger counterparts.

I wrote this book because I believe that the best innovations happen when we approach new technology with a mix of boldness and humility, courage and care. I want you to feel empowered to lead, to challenge the status quo, and to shape the future - not because technology demands it, but because your vision, your values, and your commitment to excellence make it possible.

This is my invitation to you: read these pages with curiosity, question freely, and apply what resonates. Let this be more than a book; let it be a dialogue, a guide, and, if it serves its purpose, a spark that lights your own journey in this remarkable era of change.

Welcome to the conversation. Welcome to the future - one that you have the power to shape.

George Mathew
Sydney, September 2025

THE AI HORIZON

From Sci-Fi to Business Reality

"What was once the province of Hollywood scripts and university research papers is now in the boardroom, the factory floor, the farm, the operating theatre, and the palm of your hand."

For decades, *artificial intelligence* lived in the imagination. From *2001: A Space Odyssey*'s HAL 9000 to Isaac Asimov's sentient robots, AI's image was more speculative than practical - a technology perpetually "five years away" from changing everything.

Today, that future has arrived. AI is no longer a research curiosity or a plot device. It is embedded in our phones, search engines, logistics networks, and customer service channels. It decides credit approvals, forecasts energy demand, optimises manufacturing, and even drafts legal arguments.

The pervasiveness of AI today is striking. It is no longer confined to research labs, boardrooms, or industry conferences. It has become a topic of everyday conversation, woven into the fabric of culture and society. People from every walk of life are now grappling with its promise, its risks, and its implications for the future of work, privacy, and even human identity.

A vivid example of this came when I was invited to be on a panel discussion on AI after the matinee performance of Matthew Gasda's Doomers - a play that wrestles with contemporary anxieties about technology, society, and control. It was Father's Day, and yet the theatre was full. What was remarkable was not just the attendance but the composition of the room: students, parents, professionals, artists, technologists, retirees. A true cross-section of society.

The questions that came from the audience were not academic, nor were they technical. They were deeply human:

Will AI take away my child's future job opportunities?

Can corporations use AI to monitor and manipulate us?

Who gets to decide what is fair, safe, or ethical when machines are making decisions?

The tone was one of curiosity mixed with unease. There was genuine fascination with the possibilities of AI, but also a palpable fear - fear of losing control, of becoming invisible in decision-making systems, of being reduced to data points in a machine's logic.

What this moment crystallised for me was that AI is no longer an abstract concept. It has entered the cultural bloodstream. Plays, novels, films, and art exhibitions are using it as raw material. Families are debating it at dinner tables. Employees are asking about it in workplace town halls. Citizens are questioning it in theatres. The anxiety is real, but so is the interest - and leaders cannot afford to dismiss either.

This duality of excitement and fear reflects the core challenge for businesses, policymakers, and society at large: how to harness AI's power without succumbing to its potential for control and manipulation. The conversation in that theatre was a reminder that AI is not just a technical revolution - it is a cultural one.

For business leaders - particularly in small and medium-sized enterprises - this is not a distant or optional concern. AI adoption has shifted from *strategic opportunity* to *strategic necessity*. In a business climate marked by tighter margins, accelerated customer expectations, and relentless global competition, even consumer activism, the ability to leverage AI effectively is becoming a survival skill.

This chapter explores the forces driving AI from the lab to the ledger - and why, today, the question is no longer *whether* your business should adopt AI, but *how fast* you can do it without derailing your strategy.

From Concept To Reality: A 50-Year Overnight Success

If AI feels like it "suddenly appeared," that's only because much of its development happened quietly, outside the public eye. Early AI research began in the 1950s, with pioneers like John McCarthy and Marvin Minsky envisioning machines that could "think." Those early ambitions quickly outpaced the computing power, data availability, and algorithms of the time.

By the late 1970s and 1980s, AI had entered its first *winter* funding dried up, hype dissipated, and business interest waned. But the seeds were planted.

Three factors converged in the 2010s to bring AI back into focus:

1. **Explosion of Data** - The digital economy

began producing zettabytes of structured and unstructured data.

2. **Affordable Computing Power** - Cloud platforms and GPUs dramatically lowered the cost of training complex models.
3. **Algorithmic Breakthroughs** - Deep learning techniques, particularly neural networks, began outperforming humans in narrow domains.

By the early 2020s, AI adoption had shifted from experimental to operational in many industries - though unevenly. Large technology companies led the charge, but by mid-decade, accessible cloud-based AI services opened the door for SMBs.

A Business Imperative

A persistent myth among business owners is that AI is "for big tech" or companies with billion-dollar R&D budgets. That might have been true in 2010. It is no longer true today.

Here's why:

- **Access Has Been Democratised** - Tools like ChatGPT, Midjourney, and Copilot require no data scientists on staff to deliver business value.

- **Pricing Models Have Flattened** - Many AI tools operate on affordable subscription tiers or pay-per-use, making entry points low-risk.

- **Customer Expectations Have Shifted** - AI-driven personalisation, responsiveness, and convenience have set new market baselines.

SMBs that embrace AI are seeing gains in:

- **Productivity** - Automating repetitive tasks, freeing staff for higher-value work.

- **Decision Quality** - Leveraging AI to analyse markets, customers, and supply chains faster and more accurately.

- **Customer Experience** - Personalised recommendations, proactive service, and instant query resolution.

  Mid-cap manufacturers implementing an AI-powered demand forecasting tool can reduce inventory waste by as much as 18% in its first year, freeing up capital and boosting profitability - all with an off-the-shelf solution costing less than one full-time hire.

The Golden Age of Experimentation - and the Wild West of Implementation

We are living in what could be called AI's "golden age of experimentation." Every week, new models and tools launch. APIs allow businesses to embed AI into workflows without rebuilding their systems from scratch. The barriers to testing AI are almost nonexistent.

Yet, this accessibility has created what some call the "wild west of implementation."

- Start-ups pitch half-baked AI products with flashy demos but little real-world robustness.
- Departments adopt AI tools in silos, creating fragmented and uncoordinated systems.
- Leaders rush to "AI-ify" processes without aligning them to business objectives.

The result? While AI pilots are multiplying, most fail to scale. An IDC-Lenovo research suggests that 88% of AI projects never reach production, and in generative AI, the number is even higher. Another study from MIT put the number as high as 95%. The issue is rarely the technology - it is poor *strategic alignment* and lack of change management.

When I worked as a Country Analyst with Gartner, we regularly published what became one of the most recognisable tools in the technology world: the

Gartner Hype Cycle. At its core, it's a framework for understanding how new technologies mature and how businesses should interpret the promises versus the realities of adoption.

The model is deceptively simple yet powerful. Every new technology wave travels through a "Peak of Inflated Expectations" - a period of media buzz, soaring vendor promises, and boardroom excitement. Companies rush to experiment, often with the belief that competitive advantage lies simply in being first.

But soon, the harder truths emerge: integration challenges, inflated costs, data readiness issues, and disappointing pilot outcomes. This leads many organisations into the "Trough of Disillusionment." Here, enthusiasm wanes, projects are cancelled, and headlines begin to question whether the technology was overhyped all along.

Over time, however, the dust settles. Best practices emerge, failures yield valuable lessons, and vendors refine their offerings. Those who stay the course move towards the "Slope of Enlightenment", where practical, scaled applications begin to prove lasting value. Finally, a technology reaches the "Plateau of Productivity", where adoption becomes mainstream and ROI is predictable.

So where does Artificial Intelligence sit on this

curve today? Opinions differ, but I would argue that AI currently occupies a place somewhere between the Peak of Enlightenment and the Trough of Disillusionment. The initial excitement of generative AI - those dazzling early demos of chatbots, image generators, and predictive analytics - has given way to a more sobering reality of getting them to work with complex organisational data. Many pilots have stalled, either because the data isn't ready, the ROI is unclear, or the organisational culture hasn't adapted to an AI-first mindset.

Misconceptions Holding SMBs Back

In my consulting work with SMBs across industries, I encounter three recurring misconceptions that stifle AI adoption:

1. **"It's Too Expensive"**
   Reality: Cloud AI services have dramatically reduced the cost of entry. Many tools charge per API call or seat, allowing small-scale adoption before scaling up.

2. **"It's Too Complex"**
   Reality: AI tools are increasingly no-code or low-code, with user-friendly interfaces. Complexity is more often in integration and change management than in the tools themselves.

3. **"We're Not Ready"**

Reality: Waiting until you're "ready" risks being overtaken by competitors who learn by doing. A strategic, phased approach mitigates risk without paralysis.

AI's Twin Value Proposition: Efficiency and Capability

AI's appeal lies in two intertwined promises:

- **Efficiency** - Doing what you already do, but faster, cheaper, and with fewer errors.

- **Capability** - Doing what was previously impossible, such as predicting maintenance before failure, generating bespoke product designs on demand, or analysing global market shifts overnight.

For SMBs, efficiency gains are often the first ROI driver. But capability expansion - entering new markets, developing new offerings, or serving customers in new ways - is where AI transforms a business model.

Strategic Imperative

The AI horizon is no longer distant. It is here. The leaders who will thrive today and beyond are those who:

- View AI as a strategic lever, not a side project.

- Embed AI into core processes and decision-making.
- Balance rapid experimentation with disciplined implementation.

SMBs, in particular, cannot afford to wait. The democratisation of AI means that competitors of all sizes can access similar tools. What will differentiate winners from laggards is how well they align AI with business strategy and execute at pace.

AI is both the most overhyped and underutilised technology of our era. The hype distracts from the hard, unglamorous work of aligning AI with business goals, integrating it into processes, and preparing people to use it effectively.

In the chapters ahead, we will move beyond the buzzwords to explore *how* organisations - especially SMBs - can navigate this AI horizon with clarity, discipline, and ambition. We will examine the building blocks of successful AI strategy, the pitfalls that derail adoption, and the real-world examples that show AI's transformative potential when applied with intent.

Because today, AI is no longer science fiction. It is the operating system of business reality.

THE QUIET REVOLUTION

Adapting to Tomorrow's Trends Today

"The future isn't approaching. It's already here. It's embedded in the devices we wear, attuned to our emotions, and increasingly shaping the choices that govern our daily lives."

For much of the past decade, business leaders and society at large have framed AI as a future horizon-a destination we will eventually reach. Yet the truth is more subtle and more sobering: the AI revolution is already here. It has not arrived with fireworks or sudden proclamations. Instead, it has woven itself into the fabric of daily life and enterprise, often so quietly that its presence was overlooked until ChatGPT was launched in 2022.

Think of the smartphone in your pocket. It

predicts your next word, corrects your spelling, and interprets your intent before you finish typing. Your email already filters spam with near-perfect accuracy. Your streaming service recommends the next show, not by chance, but by analysing millions of patterns across the globe. These are not incremental improvements-they are evidence of a new mode of living and working, powered by algorithms that continuously learn.

I was in Bangalore, India, with family just before Christmas when ChatGPT was launched. I remember telling everyone that this was a game changer, bigger than the internet, mobile phones and social media combined.

In that moment I knew the world had changed instinctively because at LexX Technologies we were already training GPT models to answer complex troubleshooting questions in aviation and maritime. The difference with ChatGPT was its algorithms trained on trillions of internet pages, that as we now know it would revolutionise how people worked and lived.

Just months earlier, I had deployed a LexX virtual assistant powered by GPT in four wind farms in Illinois to support technicians for troubleshooting 400 odd Vestas V82 turbines. The excitement on the ground was palpable, that wind technicians now had the entire knowledgebase required to fix a turbine in their back pockets that can be queried in natural language via a conversational chat. ChatGPT's scale and impact was unfathomable.

So the challenge is not whether AI will shape the future - it already has and continues to do so. The question is whether leaders, boards, governments, and individuals can adapt to its pace, respond to its demands, and reimagine their relevance in a world that refuses to pause for hesitation.

This chapter explores the *quiet revolution* of AI: the way it infiltrates industries, redefines human interaction, and accelerates change in forms once reserved for science fiction. It is a wake-up call for those tempted to adopt a "wait and see" approach. As we will discover, complacency is no longer a viable strategy.

The Quietness Of The Revolution

Why "quiet"?

The AI revolution has not been heralded by massive structural overhauls in most industries, but by gradual improvements embedded in everyday tools. Unlike the internet boom of the 1990s, which was impossible to ignore, AI has seeped into our workflows, services, and homes incrementally. It improves customer service wait times without fanfare. It optimises supply chains invisibly. It helps radiologists detect anomalies in scans with higher accuracy, yet the patient may never know an algorithm was involved.

This quietness is deceptive. It leads many to believe the revolution has yet to arrive, when in reality, it is shaping strategic landscapes, determining winners and losers, and rewriting rules of engagement. The audience of the play I saw on Father's day knew this. They chose to spare a Sunday afternoon discussing it. Yet on that same day, at many Father's Day lunches the topic of AI may never have surfaced.

The greatest risk for businesses is not disruption by an unseen competitor, but disruption by their own inertia. Leaders who mistake subtlety for insignificance risk being left behind by competitors who embrace the undercurrents of AI today.

Skills as the new currency

The future will not belong to the strongest incumbents or even the wealthiest firms, but to those who can learn, adapt, and reskill fastest. AI transforms not just industries but professions. Roles once considered immune to automation - lawyers, doctors, designers - are already seeing core tasks augmented or replaced by algorithms.

For individuals, the call is clear: upskill, cross-skill, or risk obsolescence. For boards and executives, the mandate is sharper: invest in human capital with the same urgency as financial capital.

Redefining relevance

Organisations must continuously redefine why they matter in an AI-driven market. Competitiveness is no longer about products or processes alone; it is about resilience, adaptability, and cultural openness to change.

Emerging Digital Trends

The following frontiers may sound futuristic, but they are already here-piloted, commercialised, and scaling.

Brain-Computer Interfaces (BCI)

Companies like Neurable and Nextmind are

pioneering direct links between human thoughts and digital systems. From controlling devices without hands to enabling communication for people with disabilities, the potential of BCI stretches across healthcare, defence, and consumer electronics.

The leadership challenge: ethical boundaries and workplace integration.

Synthetic Data-as-a-Service

Firms like Mostly AI and Synthetaic offer synthetic datasets that mimic real-world patterns without exposing private information. This revolutionises training models while addressing privacy concerns.

The leadership challenge: ensuring synthetic data still reflects real-world diversity, avoiding algorithmic blind spots.

Decentralised Physical Infrastructure Networks (DePINs)

Projects like Helium (for decentralised wireless networks) and Hivemapper (for community-driven mapping) shift infrastructure building from centralised corporations to distributed communities.

The leadership challenge: governance, monetisation models, and long-term sustainability.

Emotion AI in Customer Experience

Emotion AI interprets facial expressions, tone, and micro-gestures to tailor customer interactions in real-time. Retail, healthcare, and HR are early adopters.

The leadership challenge: ethical concerns about surveillance and emotional manipulation.

Spatial Commerce

Brands like Nike, IKEA, and Shopify are driving geo-triggered offers, AR-powered fitting rooms, and location-aware shopping. Commerce is no longer confined to stores or websites-it follows the customer wherever they are.

The leadership challenge: ensuring relevance without intrusiveness.

Digital Twins of Humans

Beyond digital twins of machines, companies now explore twins of individuals-avatars that simulate behaviour, preferences, and health conditions.

The leadership challenge: ownership of identity, privacy, and informed consent.

Zero UI

Voice, gestures, and sensors increasingly replace screens. Amazon and Samsung lead the race to design environments where technology fades into the background, and interaction becomes natural.

The leadership challenge: ensuring accessibility and inclusivity.

AI-Generated Laws

Nations like Estonia and the UAE experiment with AI-assisted legislation drafting. This may shorten bureaucratic processes but raises profound questions of democratic accountability.

The leadership challenge: preserving transparency

and public trust.

The Danger of a Laissez-Faire Mindset

A laissez-faire approach-assuming "we can adapt later"-is perilous. AI does not wait for annual strategy meetings. It compounds daily, across competitors, startups, and regulators. Those who defer action will find not just their systems outdated, but their entire business models irrelevant.

Adaptation is not optional; it is survival.

Building The Adaptive Enterprise

To thrive in this quiet revolution, organisations must embed adaptability into their DNA:

1. **Leadership Vigilance** – Boards and executives must scan emerging trends constantly.
2. **Continuous Learning** – Upskilling programmes should be as routine as compliance training.
3. **Human-AI Partnership** – Frame AI as augmentation, not replacement.
4. **Governance by Design** – Anticipate ethical dilemmas before they escalate.
5. **Experimentation Culture** – Encourage pilots, but ruthlessly scale or abandon based on results.

Strategic Imperative

For years, the AI revolution advanced quietly optimising supply chains, reducing wait times, and assisting professionals in ways that often went unnoticed. But the launch of ChatGPT in late 2022 shattered that quietness. Suddenly, AI was no longer an invisible force working behind the scenes; it became a household name, a subject of dinner-table debates, boardroom deliberations, and public imagination. What had crept in as a subtle undercurrent became a tidal wave that no leader can now ignore.

The future will not be secured by incumbency, scale, or financial strength alone. It will belong to organisations and individuals who learn, adapt, and reskill at speed. Skills have become the new currency of competitiveness, and boards must invest in human capital with the same urgency as financial capital. Adaptability, not tradition, will define resilience.

Relevance in the AI age is not about products or processes, but about an organisation's ability to continuously reinvent why it matters. The imperative for leaders is clear: act decisively in this post-ChatGPT era, or risk being swept aside by a revolution that will no longer wait quietly.

BUSTING THE MYTHS

What AI Isn't for Your Business

"Many business owners and mid cap CEOs hold onto beliefs that prevent them from leveraging AI's immense potential."

In my 25 years of working with businesses across industries, I've found that the single greatest barrier to innovation is rarely the technology itself. It's the *beliefs* about technology. AI has perhaps the most real example where the invisible barrier to adoption is real and tangible.

The hype around AI is relentless. Every news cycle brings stories of AI breakthroughs - or AI disasters. For small and medium-sized businesses (SMB), the effect is often paralysis. It's not that leaders can't see AI's potential; it's that they're unsure where to begin, afraid of making costly mistakes, or overwhelmed by jargon and complexity.

These misconceptions are sticky. They spread in conference conversations, in LinkedIn posts, and even in board meetings. Left unchallenged, they can stall decision-making for years - years in which competitors quietly move ahead.

In this chapter, we dismantle five of the most common myths holding back SMBs. These myths are not harmless misunderstandings. They are active obstacles. By replacing them with a grounded, practical understanding, business leaders can unlock new opportunities and chart a clear path to AI adoption that is *realistic*, *affordable*, and *aligned to business priorities*.

Myth 1: "AI Is Too Expensive"

When I speak at industry events, this is the myth I hear most often - and most passionately. The image many leaders have is of AI requiring multi-million-dollar data centres, armies of data scientists, and years of development before seeing any return.

That picture is outdated by at least a decade.

Reality: Affordable, scalable tools already exist. Thanks to cloud computing and the as-a-service economy, businesses can access advanced AI capabilities without owning any of the infrastructure. You can rent machine learning models by the hour, or pay a subscription for a specific function - sentiment analysis, image recognition, automated report generation - often at a fraction of a single full-time salary.

Example:

A regional freight company in New South Wales, Australia uses an off-the-shelf AI route optimisation tool costing less than $2,000 a year. It reduced fuel costs by 12% in the first three months. The AI wasn't "custom-built" - it was a subscription service integrated into existing GPS systems.

Actionable Path Forward:

1. **Start small** - Begin with a single, well-defined problem and test a low-cost AI tool to address it.

2. **Use the free tiers** - Many providers offer limited but functional free access to AI services.

3. **Leverage your existing vendor relationships** - Ask your ERP, CRM, or HR software provider about AI add-ons you already qualify for.

Myth 2: "AI Will Replace My Staff"

Headlines warning of "robots taking jobs" have been around since the industrial revolution. With AI, those headlines have returned - and they're louder. For SMB leaders, the fear is often twofold:

1. Loss of valued staff and the institutional knowledge they hold.

2. Cultural resistance from employees worried about being replaced.

Reality: AI's primary business value lies in augmentation, not replacement.

AI is best at repetitive, rules-based, high-volume tasks - the kind that consume disproportionate amounts of human time but require minimal judgement. When applied well, AI *frees* your human staff to focus on tasks that require creativity, empathy, problem-solving, and relationship-building - all areas where humans still outperform machines.

Example:

A mid-sized accounting firm in Brisbane, Australia implemented AI-driven expense categorisation. Staff who once spent hours coding transactions now spend that time advising clients on tax strategy and compliance. Not a single staff member was let go; instead, client satisfaction scores went up and revenue per client increased.

Actionable Path Forward:

- **Communicate early and often** - Explain to your team *how* AI will change their roles and why it benefits them.

- **Reinvest saved time** - Redirect productivity gains into higher-value activities.

- **Pair training with deployment** - Introduce upskilling programs alongside AI adoption to ensure staff feel part of the future, not threatened by it.

Myth 3: "I Need to Be a Tech Expert to Use AI"

A decade ago, implementing AI often meant writing Python scripts, configuring servers, and wrangling data pipelines. That era is over - at least for SMBs adopting *applied AI*.

Reality: Many AI tools are built for non-technical users. Drag-and-drop interfaces, natural language commands, and plug-and-play integrations mean you can deploy AI in a morning without writing a line of code.

Think of modern AI tools like Excel in the 1990s. You didn't need to know how to write the underlying software; you just learned enough formulas to do your job better.

Example:

A boutique marketing agency in Melbourne,

Australia uses an AI-driven ad optimisation platform. The onboarding took one hour. Campaign managers simply adjust sliders and approve AI-generated suggestions - no coding required.

Actionable Path Forward:

1. **Prioritise tools with strong UX** - Look for platforms that emphasise ease of use in their demos.

2. **Use vendor onboarding** - Many providers include free training or video tutorials.

3. **Nominate a "translator"** - Identify someone in your team who enjoys learning new tools and can help others adapt.

Myth 4: "AI Is Something I Have to Buy Separately"

This myth is particularly damaging because it hides AI's presence in the tools businesses already use daily.

Reality: You probably already have AI in your business.

If you use Microsoft Office 365, you've got AI-powered grammar suggestions, spreadsheet analysis ("Ideas" in Excel), and integrated meeting transcription in Teams. If you use Gmail, its spam filter is AI. If you use a modern CRM, chances are it offers AI-driven lead scoring.

Example:

A manufacturing SMB in Geelong, Australia discovered its inventory management software already had a machine learning module to predict stockouts. By turning it on (at no additional cost), they reduced backorders by 22% in one quarter.

Actionable Path Forward:

- **Audit your tech stack** - List every software product you use and identify built-in AI features.
- **Activate dormant features** - Many require only configuration, not new licences.
- **Measure ROI quickly** - Track the impact of using existing AI before investing in new tools.

Myth 5: "AI Is Overhyped - It's Impossible to Choose the Right Products"

Given the noise in the AI marketplace, this myth has some grounding in reality. The hype is real - and dangerous. Many vendors lead with AI branding before they have a viable product.

Reality: The solution is not to avoid AI, but to reframe the selection process.

The most effective way to cut through the noise is to start with your business problem, not the technology. AI should be an enabler, not the starting point.

Example:

A wholesale distributor in Perth, Australia wanted to reduce customer churn. They didn't go shopping for "AI products." They defined their problem: customers who hadn't ordered in 90 days were at risk. Then they searched for tools that could predict churn and trigger re-engagement campaigns. The winning solution happened to be AI-powered, but that was secondary to solving the problem.

Actionable Path Forward:

1. **Define success first** - What specific business outcome do you want?

2. **Evaluate vendors against your criteria** - Ignore features that don't directly contribute to solving your problem.

3. **Run a pilot** - Test with a small dataset or team before committing.

The Cumulative Effect of Busting Myths

When these myths are dismantled, something powerful happens: hesitation turns into momentum. Business leaders stop asking, "Can we afford AI?" and start asking, "Where should we apply it first?"

This shift in mindset is critical. AI adoption is not a binary switch; it is a series of small, strategic steps that compound over time.

Strategic Imperative

These five myths are not the only misconceptions about AI - but they are the most common and the most damaging for SMBs. Left unchallenged, they perpetuate inaction. But once replaced with the practical realities, they create a foundation for confident, targeted AI adoption.

In the next chapter, we move from *myth-busting* to *method-building*. We will explore the practical framework for selecting, piloting, and scaling AI in a way that matches your business's strategic priorities and risk appetite.

Because AI is not something to fear - or to worship. It is a tool. And like any tool, its value depends entirely on the skill and intent of the person using it and leaders that embrace it.

THE AI ICEBERG

Why Most Companies Fail to See What's Below the Surface

"Beneath the surface lies an entire world of challenges and decisions that will make or break your AI investments."

From a distance, the Dangerous illusion of simplicity is obvious AI looks like a straight line:

Step 1: Gather data.
Step 2: Apply AI.
Step 3: Create value.

It's a seductive narrative because it mirrors the way software has historically delivered business benefits. You license a system, feed it information, and expect outputs that improve efficiency or revenue.

But AI isn't just software. It's a living, learning

system - and that changes everything.

Like an iceberg, the shiny promise of AI - the part everyone sees - is only the small tip above the waterline: flashy dashboards, predictions, chatbots, and automation. The bulk of the work, risk, and strategic decision-making lies hidden below the surface:

- Preparing data that's messy, inconsistent, and fragmented across systems.
- Understanding and managing bias in ways that protect fairness, compliance, and brand reputation.
- Choosing platforms and vendors without locking yourself into costly, underperforming ecosystems.
- Embedding governance, explainability, and resilience into your AI processes.

These below-the-waterline issues determine whether your AI initiative creates lasting value - or becomes a sunk cost that erodes trust inside and outside your organisation.

The AI Iceberg Model - Above vs. Below the Waterline

Above the waterline, executives see:

- AI demos that "wow" in boardrooms.
- Promises of efficiency and predictive power.
- Vendor assurances that their platforms are turnkey.

Below the waterline lies the reality:

- 70–80% of effort in AI projects goes into data cleaning, integration, and labelling.
- Model outputs are only as good as the governance and monitoring wrapped around them.
- Vendor lock-in and poor interoperability can trap you in inflexible, costly ecosystems.
- Legal, ethical, and reputational risks increase as AI decisions touch customers, employees, and regulators.

This is not meant to scare leaders away from AI. It's meant to shift attention from the *visible* to the *material*.

Challenge 1: Data Readiness - The

Unseen Heavy Lifting

One of the most persistent misconceptions is that organisations can simply feed their existing data into an AI system and expect results.

Reality: Your data is *almost never* AI-ready.

Why AI-Ready Data Is Rare

- **Fragmentation:** Data lives in disconnected systems - CRM, ERP, spreadsheets, cloud drives, and even paper.

- **Inconsistency:** Different teams record the same type of information in different formats.

- **Quality issues:** Missing fields, duplicates, outdated records, and human entry errors can undermine AI performance.

An AI model trained on inconsistent or biased data will not just reflect those flaws - it will amplify them.

Practical Steps for Data Readiness

1. **Data Inventory:** Identify what data you have, where it resides, and in what condition.

2. **Data Cleaning:** Deduplicate, standardise formats, and fill gaps.

3. **Data Governance:** Establish ownership, update protocols, and security controls.

4. **Ethical Data Use:** Ensure data is collected and processed in compliance with relevant regulations (GDPR, CCPA, Australian Privacy Principles).

Challenge 2: Bias and Explainability - The Trust Foundation

The Workday and Mary Louis cases illustrate why bias is not an academic concern - it is a business risk with legal and reputational consequences.

The Workday Case

In 2025, a U.S. federal judge allowed a class action lawsuit filed in 2024 to proceed against Workday Inc., alleging that its AI-driven hiring tools discriminated against applicants over the age of 40, as well as against minority groups. The claim: the algorithm effectively screened out qualified candidates based on attributes that correlated with protected characteristics.

The Mary Louis Case

In another example, a company settled for USD $2.2 million after its AI-powered algorithm was found to disadvantage certain rent applications from sections of society.

These cases show:

- AI can inadvertently replicate historical human biases present in the data.
- Even if the discrimination is unintentional, the legal exposure is real.
- Public perception can shift rapidly against companies seen as "hiding behind algorithms."

Reality: Bias and explainability are not "add-ons" for later; they are core design requirements.

Practical Steps for Bias and Explainability

1. **Diverse Training Data:** Ensure datasets are representative of the populations they affect.
2. **Bias Testing Protocols:** Test outputs for disparities across demographic groups.
3. **Explainability Tools:** Use model interpretability techniques to explain why decisions were made.
4. **Human-in-the-Loop:** Keep a qualified human reviewer for high-impact decisions.

Human bias predates AI by centuries - the technology is not inherently biased. But when data with bias in them is embedded into automated systems, it scales faster and becomes harder to detect.

Challenge 3: Independent Advice - Guarding Against the Vendor Mirage

AI vendors are in the business of selling solutions - often positioned as the only viable path to success. But what you see in the sales demo may not reflect the realities of deployment, integration, or long-term scalability.

Reality: Independent advice before committing to any AI vendor or platform is advisable.

Why Independence Matters

- **Avoiding Lock-In:** Proprietary formats can make it costly to switch vendors.
- **Matching Needs to Solutions:** Not all AI features will be relevant to your core objectives.
- **Identifying Hidden Costs:** Storage, training data labelling, and compute charges can balloon.

Practical Steps for Independent Vetting

1. **Hire or Consult an Independent Expert:**

Someone not tied to a specific vendor.

2. **Run a Paid Pilot:** Test performance on your actual data before committing.

3. **Negotiate Portability:** Ensure you can export data and models if needed.

Strategic Investments Below the Waterline

The executives who succeed with AI invest strategically in the parts of the iceberg most competitors ignore:

1. **Data Infrastructure:** Building pipelines that deliver clean, timely, and relevant data.

2. **Governance:** Policies, roles, and processes for ethical and compliant AI use.

3. **Capability Building:** Training staff to work with AI outputs and challenge them when necessary.

4. **Resilience Planning:** Designing systems that fail gracefully and recover quickly.

These investments don't produce immediate "wow" moments in the boardroom. But they produce *durable* AI value - value that isn't wiped out by a lawsuit, a compliance investigation, or a data breach.

The Executive Mindset Shift

Above-the-waterline thinking focuses on *acquisition*: "What tool should we buy?"
Below-the-waterline thinking focuses on *foundation*: "What capabilities must we build so any AI tool delivers value safely and sustainably?"

The shift requires patience and a tolerance for invisible progress. Cleaning data, building governance frameworks, and running bias audits don't create instant shareholder buzz - but they create the conditions for AI success that lasts.

Strategic Imperative

The AI iceberg analogy is more than a metaphor; it's a reminder that the unseen determines the fate of the visible. The executives who invest below the waterline will not just avoid costly failures - they will position their companies for resilient, compounding returns from AI over years, not months.

In the next few chapters, we will move from *understanding* the iceberg to *navigating* it - outlining a practical playbook for designing AI initiatives that surface value without capsizing under unseen risks.

Because AI done well is not about rushing to the tip of the iceberg. It's about shaping the entire structure so it can support your business through shifting waters.

THE ART OF FRAMING THE RIGHT PROBLEM

Before the Techies Get Involved

"The greatest value is created not when you start solving problems, but when you obsess about defining the right problem."

In today's corporate climate, speed has become a badge of honour. Senior leaders are often praised for their ability to "move quickly" and "get things done". Yet in the race to adopt new technologies, particularly AI, speed without direction can be fatal. The rush to find solutions before truly understanding the underlying problem is one of the most common reasons digital transformation efforts fall short. The seduction of the solution mode is real.

It begins innocently enough. A board member reads about a competitor implementing AI chatbots. A CIO attends a conference where predictive maintenance is touted as a game-changer. A department head sees a demo of an AI analytics dashboard and believes it could solve their reporting headaches. Before long, vendors are in the building, proposals are being written, agile squads are spun up, and money is committed - all before anyone has paused to ask: *What exactly are we trying to solve?*

The premature leap into "solution mode" is seductive. It creates a sense of momentum and vague purpose. It reassures stakeholders that progress is being made. But beneath this apparent activity lies a dangerous truth: without a precise understanding of the problem, technology projects are essentially experiments in hope.

Over the years, I have seen this pattern play out in industries as diverse as aviation, financial services, and renewable energy. A large bank once commissioned a machine learning project to improve loan approvals, only to discover months later that the real bottleneck was an outdated manual verification process - something that could have been fixed in weeks with process redesign. A wind farm invested in predictive analytics to optimise turbine performance, but the biggest losses were occurring from a poorly coordinated maintenance schedule. In both cases, the problem framing was flawed from the outset, and millions were wasted chasing solutions to the wrong problem.

This chapter is about resisting the urge to outsource the thinking to vendors or jump to solutions too quickly. It is about cultivating the discipline to slow down, ask better questions, and truly understand the problem before you let the technologists loose. Because when you get the problem framing right, the path to breakthrough innovation becomes clearer, and the solutions that emerge tend to be not only effective but also sustainable.

The Cost of Solving the Wrong Problem

Solving the wrong problem is not just inefficient - it is actively harmful. The costs extend beyond wasted budget.

First, there is the **opportunity cost**. Every hour, dollar, and brain cell spent chasing the wrong problem is time and energy that could have been directed towards something truly transformative. In competitive industries, these lost opportunities can mean missing entire market windows.

Second, there is the **erosion of trust**. When a high-profile AI initiative fails to deliver, stakeholders become sceptical of future projects. This scepticism spreads through the organisation, making it harder to secure buy-in for genuinely valuable initiatives later.

Third, there is the **cultural damage**. Teams who have invested their energy into an initiative that fails due to poor problem framing often feel frustrated and disengaged. They may be less willing to contribute their insights to future projects, believing their input will be ignored in favour of pre-decided solutions.

I recall working with a maritime logistics company that spent 18 months and significant capital building an AI-based scheduling system. The system was technically impressive but failed to account for a simple reality: the biggest delays were caused by weather-related port closures, not scheduling inefficiencies. By the time this was realised, the project's credibility was beyond repair.

The lesson here is clear. If you cannot clearly define the problem, you cannot meaningfully evaluate whether a proposed solution is appropriate. Problem framing is not optional. It is foundational.

The Better Way: Ruthless and Empathetic Problem Framing

Effective problem framing requires a deliberate mindset shift. Instead of measuring success by how quickly you can get a project into delivery, you measure success by how deeply you understand the problem before committing to a solution.

This approach feels counterintuitive in fast-moving corporate environments. Leaders worry that slowing down will make them appear indecisive. In reality, it positions them as disciplined and strategic. The paradox is that the time spent framing the problem correctly often results in faster, more successful delivery downstream.

The most effective leaders I have worked with treat problem framing as a **non-negotiable** phase in any innovation initiative. They are willing to say "no" to premature solutioning and insist on clarity before action. This is not about creating bureaucratic hurdles. It is about ensuring that every subsequent dollar and decision is anchored in reality.

Below are four practical steps to master the art of problem framing.

Step One: Walk the Floor. Ride the Truck. Live the Day.

You cannot frame a problem effectively from behind a desk. You have to see it, feel it, and hear it in its natural environment.

This means spending time with the people who live with the problem every day - the call centre agent who has to deal with irate customers, the field technician who spends hours driving between job sites, the analyst who spends half their week fixing data errors.

In maintenance, I have joined engineers in workshops, watching them troubleshoot faults on grounded assets. The insights gained in those hours were far richer than anything that came from formal status reports. In the renewable energy sector, riding in a service truck with a wind turbine maintenance crew revealed operational bottlenecks that no spreadsheet could have highlighted.

This immersion is not just about observation. It is about empathy. When people see that you are willing to step into their world, they open up in ways they rarely do in meeting rooms. They point out the small frustrations, the workarounds, the things they have learned to tolerate. Often, these are the clues to the real problem.

During my time at Cochlear, we undertook the ambitious task of rolling out a global sales and marketing platform that would ultimately impact operations across 32 countries. The scale of the initiative was immense: we were not just implementing technology, but also transforming how sales teams, marketing functions, and customer engagement processes aligned in markets as diverse as Europe, Asia, and the Americas.

One of the most critical lessons from that program was the discipline of listening before solving. Rather than dictating requirements from headquarters, we deliberately designed a process to capture the voices of local markets. To do this, we sent facilitators and subject matter experts to seven regional locations. Their mission was not to evangelise the platform or prescribe solutions, but to gather the insights of senior executives and local leaders who carried deep knowledge of their markets - knowledge of regulatory environments, customer needs, cultural nuances, and healthcare pathways - to frame the problem.

The facilitators were strictly briefed not to discuss solutions. This was non-negotiable. The moment a facilitator hints at a solution, the conversation narrows; participants stop sharing openly and begin reacting to what they believe is being proposed. By stripping away that bias, we created a safe environment where local executives could articulate pain points, opportunities, and unique constraints without fear of being overruled by a global "one-size-fits-all" approach.

The result was a body of insight that shaped not only the design of the platform but also its adoption. The diversity of perspectives forced us to confront uncomfortable truths: what worked brilliantly in one market could fail spectacularly in another due to regulatory restrictions or differences in healthcare systems. Instead of resisting this complexity, we leaned into it, building flexibility into the platform and enabling local teams to adapt global capabilities to their specific contexts.

This experience serves as a powerful reminder for any organisation embarking on an AI transformation. Technology cannot be imposed in isolation; it must be informed by the voices of those who live closest to the market reality. The discipline of listening before prescribing solutions is not simply good stakeholder management - it is the foundation of sustainable change.

Step Two: Sift Through the Noise – Identify High-Value Problems

Once you have gathered observations, the next challenge is to separate signal from noise. Not every frustration or inefficiency is worth solving.

High-value problems tend to share certain characteristics:

- **Frequency:** They occur often enough to have a cumulative impact.
- **Cost:** They result in significant financial, operational, or reputational damage.
- **Emotional Drain:** They cause frustration, stress, or disengagement among staff or customers.
- **Value Unlock:** Solving them opens the door to broader improvements or competitive

advantage.

In one project with a large utility provider, we discovered that the single most impactful problem was not the inefficiency of field operations, as originally assumed, but the delay in customer communication after outages. Addressing this issue improved customer satisfaction scores far more than any operational optimisation would have.

By applying these criteria, you avoid the trap of diluting your efforts across dozens of minor issues.

Step Three: Prioritise Ruthlessly – Focus Beats Consensus

Organisations often fall into the trap of trying to please everyone by tackling multiple problems at once. The result is diluted focus and mediocre outcomes.

Ruthless prioritisation means making hard choices. You focus on the one or two problems that will deliver disproportionate impact. This requires courage, because it often means saying "no" to influential stakeholders whose pet issues did not make the cut.

In a large manufacturing company, we narrowed a list of 17 identified problems down to two. These two, when solved, addressed root causes that were driving many of the other issues. The initial pushback was intense, but the eventual results vindicated the decision.

Step Four: Define the Problem – Then Bring in the Builders

Only after you have walked the floor, sifted through the noise, and prioritised the high-value problems should you move into defining the problem statement.

A good problem statement is:

- Clear and unambiguous
- Anchored in evidence
- Free of implied solutions

Let us look at some examples

Example 1

Over the past 12 months, customer complaints about failed international transactions have increased by 37%, with an average resolution time of 5.2 days, compared to the industry benchmark of 1.8 days. These failures primarily affect customers in Asia-Pacific corridors and account for $12.3M in potential lost revenue annually, creating reputational risk and contributing to a 6% decline in Net Promoter Score (NPS).

Example 2

Customer wait times for dinner service at our busiest location have risen from an average of 9 minutes to 18 minutes over the past quarter, while customer satisfaction scores dropped by 21% in the same period. This location represents 35% of total revenue, but turnover among kitchen staff has reached 28% annually, compared to the company average of 12%.

Example 3

The current onboarding process for new customers requires an average of 6.5 hours of manual configuration by engineering staff per client, creating a backlog of 27 customers waiting for implementation. This backlog has extended average onboarding time to 21 days, compared to our target of 7 days, and is delaying the conversion of $150,000 in pending contracts.

Once the problem is defined, you can engage technology teams or external vendors. But instead of asking them to "solve" in a vacuum, you are giving them a well-framed context. This dramatically increases the odds that the solutions they propose will be relevant and effective.

The Empathy-Analytics Balance

Problem framing is both an art and a science. The art comes from empathy - truly understanding the human dimension of the problem. The science comes from analytical discipline - ensuring that your understanding is backed by data and evidence.

Neglect either side and you risk failure. Too much empathy without evidence can lead to solutions based on anecdotes. Too much analysis without empathy can lead to technically elegant solutions that fail in practice because they ignore human realities.

Conviction to Slow Down Before Speeding Up

In high-pressure environments, it takes conviction to insist on slowing down. I am not advocating for laissez-faire attitude which is an attitude many SMB leaders have unfortunately taken. Stakeholders must push for quick wins, but a disciplined leader holds the line. This is not about delaying progress for its own sake. It is about ensuring that when you do move, you move in the right direction.

Some of the most successful AI initiatives I have seen were those where the leaders resisted pressure to jump to solutions, even when it meant difficult conversations with boards or executive teams.

Strategic Imperative

Framing the right problem is not a peripheral skill. It is a central leadership discipline in the age of AI. By investing the time to understand the real issue before engaging the technologists, you dramatically improve your chances of delivering solutions that work, last, and create real value.

In the chapters that follow, we will explore how this discipline integrates into the broader lifecycle of AI adoption, and why the organisations that master it are the ones that consistently turn technological potential into business reality.

PICKING THE RIGHT AI SOLUTION

Automation, Augmentation, or Agents?

"Not all AI is created equal-and neither are the problems we're trying to solve."

The current AI landscape is a dazzling marketplace of possibilities. Every week, a new platform, model, or framework is launched, each promising to revolutionise workflows, supercharge productivity, or outthink competitors. In this environment, it is easy - almost irresistible - to leap into solution mode with enthusiasm.

Yet beneath the shiny veneer of innovation lies a sobering truth: most AI initiatives that fail do so not because of the technology's limitations, but because of a poor fit between the chosen solution and the actual problem at hand. In other words, the wrong kind of AI is often applied to the wrong kind of challenge.

The purpose of this chapter is to slow down that decision-making process and offer leaders a structured way to choose the right *type* of AI solution. This is not simply a matter of comparing features or costs; it is about aligning the underlying *mode* of AI implementation to the strategic, operational, and cultural realities of your organisation.

We will explore the three primary modes of AI in business contexts-Automation, AI-Augmented Workflows, and AI-Driven Agents-each with its strengths, weaknesses, and ideal use cases. Alongside, we will introduce six strategic dimensions that serve as a compass when navigating these choices. By the end of this chapter, you should have a practical lens through which to evaluate not only *if* AI should be applied, but *how* it should be applied to create meaningful and sustainable value.

The Three Modes of AI Implementation

When stripped of marketing hype and technical jargon, AI deployments in organisations tend to fall into one of three modes. Understanding these modes is the first step towards making an informed decision.

Automation - "Set it and forget it"

Automation is the most mature and widely understood application of AI-like systems. Strictly speaking, not all automation is "true AI"-many implementations rely on deterministic, rule-based processes rather than machine learning.

Automation is best suited to tasks that are:

- Highly structured and predictable.
- Governed by clear, unchanging rules.
- Repetitive and high-volume.

Example use cases: For a bank automating compliance reporting automation is the clear fit. In other scenarios there are invoice processing, data entry, and robotic process automation (RPA) for standardised workflows. For instance, apply a late fee automatically after a set period of time and send a payment reminder automatically at 7, 14 and 30 days overdue. Staff only intervene if there are exceptions (e.g payment disputes).

Take another example in the retail context. You could implement a fully automated inventory management system that tracks real-time sales data from POS and e-commerce platforms, Uses historical trends and seasonal forecasting algorithms to predict stock needs by SKU, auto-generates purchase orders to suppliers when inventory levels fall below forecasted thresholds. Integrates with supplier systems to schedule deliveries automatically and update expected restock dates in the POS system. Sends alerts only for exceptions, like supply chain delays or anomalies in demand spikes.

Advantages:

- High efficiency and reliability once set up.
- Minimal need for human intervention.
- Clear return on investment when applied to labour-intensive processes.

Pitfalls:

- Brittle in fast-changing environments.
- Expensive to reconfigure if rules or data structures change frequently.
- Can entrench inefficient processes if the underlying workflow is poorly designed.

For example, A retail SMB runs an automated inventory reordering and forecasting system that has worked flawlessly for years. The system relies on fixed supplier contracts, stable tariffs, and predictable seasonal trends.

Suddenly, the government imposes new import tariffs on key product categories, while one supplier restructures its pricing model. The automation rules - built around historic pricing, shipping lead times, and reorder thresholds -are now outdated.

What Happens:

- The system continues placing purchase orders assuming old tariffs, leading to a 25% increase in landed costs unnoticed for weeks.
- Forecasting models trained on historical patterns fail to account for rapid shifts in consumer demand caused by price changes.
- Adjusting the rules and retraining models takes weeks of developer time, incurring unexpected consulting costs.
- The SMB could end up over-ordering expensive stock while cash flow tightens.

AI-Augmented Workflows – "The system helps, but humans stay in the loop"

This mode of AI is designed not to replace human effort, but to amplify it. AI-augmented workflows keep human expertise and judgement at the centre, while using AI to streamline, inform, and accelerate parts of the process. Augmentation is particularly valuable where:

- The task requires complex decision-making.
- Emotional intelligence, empathy, or contextual nuance are essential.
- Mistakes are costly, but total automation is not viable.

Example use cases: customer service triage, medical diagnostics, fraud detection, and legal case review. For example in Financial Services Fraud Detection in Real-Time Transactions the workflow would be that AI models scan thousands of daily transactions in real-time for unusual spending behaviour, using advanced anomaly detection and behavioural biometrics. Flagged cases are sent to human fraud analysts with AI-summarised insights, highlighting the reason for suspicion and relevant customer history. Analysts make the final decision on whether to freeze the card, escalate, or clear the transaction.

Why It's Augmentation:

- AI handles volume and pattern recognition.
- Human expertise ensures **false positives** don't lead to angry customers.
- Total automation would be too risky because

mistakes damage trust.

In the telecommunications sector, AI-augmented workflows can dramatically improve customer service without sacrificing empathy. AI systems scan incoming complaints from multiple channels - chat, email, and social media - to identify sentiment, urgency, and probable root causes. The system drafts responses and troubleshooting guides for simple cases, while escalating complex or sensitive complaints, such as those from vulnerable customers or regulatory disputes, to senior service representatives. These human agents validate the AI's recommendations and bring the emotional intelligence required to resolve difficult situations. This partnership allows AI to reduce handling time and ensure cases are prioritised correctly, while humans focus on solving nuanced problems with care and empathy.

Advantages:

- Combines machine efficiency with human judgement.

- Supports knowledge workers rather than displacing them.

- Can evolve over time as both humans and AI learn from each other.

Pitfalls:

- Over-reliance on AI suggestions can lead to skill atrophy, i.e. Humans can lose competence if they overly rely on AI suggestions.

- Requires robust change management and training to integrate effectively.

- May be slower than pure automation, depending on process design. AI-assisted workflows may be slower than pure automation because humans still verify, adjust, and contextualise outputs.

For example, A mid-sized law firm introduces AI to scan contracts and case documents for relevant clauses, risks, and precedents. Over-reliance on the AI leads some junior lawyers to lose confidence and practice in critical document analysis, reducing their ability to spot subtle legal nuances. Implementing the system required extensive training and ongoing change management, including workflows for verifying AI insights and mentoring on exceptions. Because every AI-generated summary must be carefully checked and contextualised, the overall review process can sometimes be slower than manual review for small batches of documents, though it scales better for large volumes.

AI-Driven Agents – "We'll tell it where to

go. It figures out how to get there."

AI-driven agents are autonomous or semi-autonomous systems capable of operating in dynamic, unpredictable environments. They are not limited to following pre-set rules; they can adapt strategies based on new data, changing conditions, and evolving objectives.

These systems excel in contexts where:

- Goals are clear, but the path to achieving them is variable.
- Environments are complex and fast-changing.
- Human monitoring is available for governance and oversight.

Example use cases: For logistics company deploying AI agents for real-time routing agents can maximise adaptability. A doctor's clinic using AI to suggest but not decide treatment plans, augmentation ensures safety and nuance. In other scenarios, there are supply chain optimisation, dynamic pricing engines, personalised marketing at scale, and autonomous navigation in controlled environments.

For example, a regional food distributor employs an AI agent to manage deliveries, inventory allocation, and supplier coordination. The system continuously analyses real-time data - traffic patterns, weather forecasts, perishable inventory levels, and supplier availability - to optimise delivery routes and stock placement. While the distributor sets overarching goals, such as minimising waste and maximising on-time deliveries, the AI adapts routes and stock levels in response to unexpected events, such as road closures or supplier delays. Human supervisors maintain oversight for compliance and governance, stepping in only when anomalies or high-impact disruptions occur. This approach reduces waste, improves delivery performance, and allows staff to focus on customer service rather than micromanaging logistics.

Take another example. A mid-sized online retailer uses an AI-driven pricing agent to optimise product prices across multiple categories in real time. The agent ingests competitor pricing data, inventory levels, demand trends, and seasonal patterns to adjust prices dynamically, aiming to maximise revenue and profit margins. The environment is fast-changing: competitors run flash sales, shipping delays occur, and customer demand fluctuates unpredictably. Human managers monitor dashboards to ensure pricing changes align with brand strategy and regulatory constraints, but the agent autonomously tests and implements new pricing strategies within those boundaries. This allows the business to react far faster than manual adjustments, capturing opportunities while freeing staff from constant monitoring.

In both these examples, the Characteristics Across SMB Applications are

1. Goals are clear, but the path is variable and data-driven.

2. Environments are dynamic and unpredictable, requiring real-time adaptation.

3. Humans provide oversight, governance, and exception handling, not constant control.

4. AI agents allow SMBs to scale operations, respond faster, and make data-driven decisions without dramatically increasing staff workloads.

Advantages:

- Highly adaptive and flexible.

- Can uncover solutions humans may not have considered.

- Scales effectively once guardrails are in place.

Pitfalls:

- Higher complexity and implementation risk.
- Requires robust governance to prevent unintended outcomes.
- Can be resource-intensive to monitor and maintain.

In our examples above, while a dynamic pricing AI agent can dramatically improve revenue, it introduces high implementation complexity. Integrating multiple data sources - competitor prices, inventory levels, customer demand patterns -requires sophisticated engineering and constant data validation. Without robust governance, the AI could inadvertently apply pricing that damages the brand, such as extreme discounts or sudden increases that trigger customer complaints or violate pricing regulations. Even after deployment, the system remains resource-intensive to monitor, as marketing managers and finance staff must continuously review dashboards, audit changes, and intervene when the AI experiments with unusual strategies that might backfire.

For the regional good distributor, an AI agent that dynamically manages deliveries and inventory introduces significant complexity, particularly when factoring in real-time traffic, perishable goods, and supplier variability. Incorrect modelling or misconfigured parameters could lead to stockouts, wasted inventory, or late deliveries, highlighting the high implementation and operational risk. To avoid unintended outcomes, such as routing a delivery through unsafe roads or violating contractual obligations with suppliers, the system requires robust governance, including clear escalation rules and oversight mechanisms. Monitoring and maintenance are resource-intensive, as staff must track agent decisions, validate exceptions, and recalibrate models when conditions change, such as seasonal demand spikes or new supplier contracts.

Six Strategic Dimensions for Choosing the Right AI Mode

Given these advantages and pitfalls selecting the right AI mode is not simply a matter of preference; it requires structured evaluation along six strategic dimensions.

Flexibility vs. Control

How much adaptability does the solution need? Automation offers predictability but little flexibility. AI-driven agents deliver adaptability but at the cost of more oversight. AI-augmented workflows sit in between, balancing both. In set-it and forget-it mode, humans have absolute control and flexibility in setting a late fee automatically after a set period of time and sending a payment reminder automatically at set intervals of days overdue. Whereas in the food distributor example the system takes complete control of optimising delivery routes and stock placement with humans stepping in only when anomalies or high-impact disruptions occur.

Structure of Data

Is the data clean, consistent, and structured, or messy and unstructured? Automation thrives on clean data. AI-augmented workflows can work with a blend. AI-driven agents often need diverse data inputs to learn effectively. Unstructured and diversity of data also injects risks as they have a high degree of variability to influence prescriptions.

Reliability vs. Adaptability

Does the solution need to be rock-solid in producing consistent results, or does it need to adapt in real time? Regulatory compliance processes demand reliability; competitive market strategies may reward adaptability.

Human Oversight

What level of human involvement is needed for safety, compliance, or brand protection? Automation minimises human touch. Augmentation keeps people in control. Agents require governance frameworks.

Risk Tolerance

What is the organisation's appetite for risk? Automation offers low-risk, proven returns. Agents can deliver breakthroughs but introduce higher levels of uncertainty.

Problem-Solution Fit

Is the chosen AI mode truly solving the defined problem, or is it simply a technology in search of a use case? This is where alignment with the problem-framing process from Chapter 4 becomes critical.

Practical Implementations

- A small accounting firm automates recurring invoice generation, payment reminders, and basic reconciliations. In terms of flexibility versus control, automation offers high control but limited adaptability—it executes pre-defined rules consistently but cannot respond to unusual invoice types or client-specific arrangements. The structure of data is critical: invoices and payments must be clean and well-formatted, because inconsistencies can trigger errors. Automation delivers high reliability, producing consistent outcomes with minimal variation, but offers little adaptability if clients change terms unexpectedly. Human oversight is minimal, mostly limited to reviewing exceptions or errors, and the risk tolerance is low since mistakes are generally easy to detect and correct. Finally, the problem-solution fit is strong: automation effectively addresses repetitive, rule-based tasks, freeing staff to focus on higher-value work without adding complexity.

A small retail store automates the reordering of popular SKUs once inventory falls below predefined thresholds. Regarding flexibility versus control, the system provides predictable control but limited adaptability, meaning it cannot respond to sudden surges in demand or supplier delays. Clean, structured data is essential; accurate stock counts and lead times ensure the system functions correctly. Automation offers high reliability under normal conditions, though unusual demand patterns require human intervention. Human oversight is low, with staff mainly managing exceptional cases such as promotions or delivery disruptions. The risk tolerance is low to moderate, as minor miscalculations can affect sales or cash flow but are manageable. The problem-solution fit is strong because the system addresses a repetitive, predictable task efficiently, freeing staff to focus on customer service.

- **AI-Augmented Customer Service Workflows – SMB Examples**

In the boutique e-commerce store, AI-assisted customer service helps triage messages and draft responses, leaving humans in control of nuanced cases. Here, Flexibility vs. Control is a central consideration. The solution needs some adaptability to handle unexpected customer queries, but human agents maintain ultimate control over sensitive interactions. Regarding the Structure of Data, the system operates on a mix of structured data (order histories, shipment statuses) and unstructured data (emails, social media messages), which makes AI-augmentation ideal since it can combine both types for decision support. Reliability vs. Adaptability is another key dimension: the solution must reliably identify urgent complaints while remaining flexible enough to adapt to new trends, such as a viral social media post affecting orders. Human Oversight is critical for brand protection and empathy, as AI suggestions alone cannot manage tone or urgency perfectly. The store's risk tolerance is moderate; misclassifying a complaint can hurt customer loyalty, so fully autonomous AI would be inappropriate. Finally, Problem-Solution Fit is strong: AI augments repetitive work while preserving the human touch, directly addressing the problem of workload overload without sacrificing service quality.

- **AI-Driven Agents**

Take the regional food distributor using an AI agent for supply chain optimisation. Here, Flexibility vs. Control leans heavily toward adaptability: the agent must reroute deliveries in response to traffic, weather, or supplier delays, which pure automation could not manage. Because the agent draws on multiple inputs—perishable inventory data, supplier schedules, traffic forecasts—the Structure of Data dimension is critical: diverse, sometimes messy data must be handled correctly for the AI to function effectively.

- Reliability vs. Adaptability presents a tension. While consistent deliveries are critical, the environment is fast-changing, so adaptability is more important than rigid reliability. Human Oversight remains necessary for governance: staff review exceptions, approve high-value changes, and monitor performance metrics to prevent unintended consequences. The company's risk tolerance is moderate to high; the agent can achieve efficiency breakthroughs but mistakes can be costly, such as stock spoilage or missed deliveries. The Problem-Solution Fit is clear: the AI-driven agent addresses the operational complexity that manual processes cannot scale to handle efficiently.
- Similarly, in the SMB e-commerce dynamic pricing example, Flexibility vs. Control is again vital. The AI agent can adapt pricing strategies in real time, but pricing managers retain oversight to ensure compliance and protect brand reputation. The agent's effectiveness depends on diverse and structured data

inputs, including competitor prices, inventory levels, and historical sales trends. Balancing Reliability vs. Adaptability is challenging: the agent must react quickly to market changes but avoid erratic pricing behaviour. Here, Human Oversight ensures that aggressive price adjustments do not harm margins or customer trust. The risk tolerance is relatively high —management is willing to experiment for revenue gains—but the Problem-Solution Fit must be regularly evaluated to ensure the AI is optimising the intended business metrics, not just executing automated adjustments.

Avoiding Common Pitfalls

1. **Shiny object syndrome:** Choosing the most advanced AI because it is new, not because it fits the problem.

2. **Over-automation:** Replacing human judgement in areas where nuance is critical.

3. **Neglecting governance:** Especially dangerous with AI agents that can act autonomously.

Implementation Pathways

- Start small, with a single high-value use case.

- Test the AI mode in a controlled environment.
- Build feedback loops and adjust mode selection if the initial choice underperforms.

Strategic Imperative

Choosing the right AI solution is not about chasing the most sophisticated technology - it is about making a deliberate, strategic match between the nature of the problem and the strengths of the solution. Automation, augmentation, and AI-driven agents each have their place, but their value emerges only when deployed with precision, clarity, and a deep understanding of organisational context.

When leaders resist the urge to rush, and instead apply disciplined thinking to AI mode selection, they not only increase the likelihood of success but also ensure that the technology serves the business, not the other way around. The real competitive advantage lies not in having the most AI, but in having the *right* AI, applied in the *right* way, at the *right* time.

AVOIDING PILOT PURGATORY

From Experimentation to Impact

> *"88% of AI pilots never make it to production. And when it comes to Generative AI. Over 70% of organisations haven't scaled more than a third of their pilots."*

It was the largest meeting room in the headquarters. The glass walls offered a sweeping view of the city, but all eyes were fixed on the presentation slide glowing on the screen: *"GenAI Customer Service Pilot Results – Q2 Update."*

The Head of AI was speaking with the fervour of someone who believed they had just cracked the company's future. Charts showed reduced handling times, improved first-contact resolution, and even a few heartfelt customer testimonials.

The CEO nodded, impressed. The CFO adjusted his glasses and asked, “So... when do we roll this out across the business?”

Silence. The Head of AI shifted uncomfortably. “Well... we still need to iron out some data integration issues. And compliance has some concerns. And we need more budget to scale.”

Six months later, the pilot was quietly archived. The staff went back to the old processes. The lesson was never officially documented, but it was clear to everyone: another promising AI project had vanished into what the team jokingly called “The AI Bermuda Triangle.”

This is *pilot purgatory*. It’s not about having a bad idea or poor technology. It’s about getting stuck in the no-man’s-land between proof-of-concept and enterprise-wide impact.

Understanding the "Pilot Purgatory" Problem

AI pilots are meant to be stepping stones to transformation. Yet, statistics paint a sobering picture:

- 88% of AI pilots never make it to production according to an IDC-Lenovo study.
- For Generative AI, the success rate is slightly better but still dire-over 70% of organisations

haven't scaled more than a third of their pilots.

A more recent MIT Media Lab study from Project NANDA re
leased a report that found that 95% of investments in Genera
tive AI have produced zero returns. This is despite 90% of
employees are using AI tools in their organisations for per
sonal productivity.

On paper, pilots make sense. They reduce risk, allow for experimentation, and provide a safe environment to test hypotheses before committing major resources. But in practice, many organisations find themselves unable-or unwilling-to move beyond them.

The result? Pilots become more like academic exercises than precursors to transformation. Teams can point to a flurry of activity, but the business impact remains invisible to customers, shareholders, and employees.

Why Pilots Stall: The Real Culprits

The failure to scale AI pilots rarely stems from technical limitations. More often, it's organisational DNA that determines whether a project thrives or fizzles out.

Lack of Strategic Alignment

When an AI pilot is initiated without a direct and measurable link to strategic business goals, it's effectively floating without an anchor. Leaders might greenlight pilots simply to "explore" AI or because a competitor is doing it, but without a clear purpose, the motivation to scale dissipates.

Siloed Ownership

A common trap is leaving the AI pilot entirely in the hands of a single department or a centralised Centre of Excellence (CoE). While these groups bring expertise, they can inadvertently isolate the initiative from the operational realities of the business. When the CoE hands the project back to the business, it can feel like an alien object-technically sound, but operationally foreign.

Resistance to Change

Even when the pilot delivers measurable results, scaling requires changes to processes, systems, and in some cases, people's roles. These shifts can trigger fear and resistance. Without strong change leadership, an otherwise successful pilot can encounter organisational antibodies that reject it.

Solving the Wrong Problem

Sometimes, the pilot addresses a problem that exists in theory but is not perceived as urgent by the business. Scaling a solution that nobody is asking for - or that doesn't solve a high-value pain point - is an uphill battle.

The Psychology Of Pilot Purgatory

The reluctance to move beyond pilots often has less to do with tangible constraints and more to do with human factors:

- Loss Aversion – Leaders fear that scaling will expose flaws that a pilot could keep hidden.

- Comfort in Experimentation – Pilots create a bubble where the stakes feel low. Scaling raises the stakes significantly.

- Over-Validation – Teams become addicted to testing and retesting, chasing perfect certainty before moving forward.

Pragmatic Steps To Bridge The Adoption Chasm

Moving from experimentation to enterprise impact requires both strategic and operational shifts.

Start with Small, Manageable Pilots That Solve High-Value Problems

Pilots should be scoped to deliver a measurable business outcome within a short time frame -

ideally 8 to 12 weeks. The focus should be on problems that:

- Are highly visible to leadership.
- Directly impact revenue, cost, or risk.
- Have a clear owner who is invested in the solution's success.

Build a Clear Scaling Path Before the Pilot Begins

Before launching, define the exact conditions under which the pilot will be deemed successful and ready for scaling. This includes:

- Target metrics.
- Budget approvals.
- Technical dependencies.
- Change management plans.

Integrate Business and Technical Ownership

The pilot team should be cross-functional, combining data architects (and scientists), engineers, and operational leaders from day one. This ensures the solution is both technically viable and operationally relevant.

Address Organisational Resistance Early

Anticipate process changes, role shifts, and new skills that will be required at scale. Begin engaging affected teams during the pilot, not after. This builds trust and reduces the shock of adoption.

Gradual Scaling: From Beachhead to Enterprise

Instead of attempting a full-scale rollout immediately, scale in stages:

1. Beachhead – Deploy in one department or market.

2. Expansion – Add additional departments or geographies.

3. Enterprise – Full organisational adoption.

This phased approach reduces risk while building momentum.

Avoid the "AI Pilot Graveyard"

Maintain a central repository of all pilots-successful and unsuccessful - with documented learnings. This prevents future teams from unknowingly repeating failed approaches and helps leadership see patterns over time.

The Role of Executive Sponsorship

An AI pilot without an engaged, senior-level sponsor is almost guaranteed to stall. The sponsor's role is not simply to approve budgets-it's to:

- Shield the project from shifting political winds.
- Connect it to broader strategic priorities.
- Champion scaling when success criteria are met.

Without this advocacy, even the most technically brilliant pilot risks becoming another abandoned artefact in the archives.

Metrics That Matter in Scaling Decisions

To break free from pilot purgatory, leaders must demand the right metrics-not just technical performance, but indicators of business impact:

- Return on Investment (ROI) – Projected and actual.
- Time to Value – How quickly benefits are realised.
- Adoption Rates – Are end users actually using it?

- Operational Fit – Does it integrate cleanly into existing workflows?

Strategic Imperative

Pilot purgatory is not a sign that AI is overhyped or unready - it's a sign that organisations are still learning how to bridge the gap between possibility and production.

Escaping this trap requires clarity of purpose, cross-functional ownership, and a willingness to face the messy realities of organisational change. The organisations that succeed are those that treat pilots not as safe little science experiments, but as the first step in a disciplined march toward impact.

In AI, the goal is not to run more pilots - it's to run fewer, better pilots, and to have the courage and conviction to scale the ones that work.

AI FOR SMBS

Start Smart, Spend Little, Scale Wisely

"AI isn't just for big tech or billion-dollar budgets. Today, smart AI adoption is becoming a survival skill for small and medium-sized businesses."

For many small and medium-sized businesses, AI feels like a distant luxury. The common perception is that AI is only accessible to the Amazons, Apples, and Googles of the world. As large corporations would, the idea of investing in data science teams, large-scale automation, and custom AI models seems financially and technically unattainable. Add to this the endless parade of AI buzzwords, conflicting vendor claims, and daily headlines about breakthroughs that seem light years away from everyday business, and it is no wonder many SMB leaders hesitate.

Yet the reality today is strikingly different. Smart AI adoption is no longer an exotic innovation; it has become a survival skill. Competitors, regardless of size, are finding ways to use AI to operate leaner, serve customers faster, and make better decisions with less guesswork. The beauty of AI today is that its entry point for SMBs has never been lower. Freemium tools, no-code automation platforms, cloud-based AI services, and an ecosystem of affordable expertise have flattened the playing field.

This chapter offers a pragmatic roadmap for SMBs to approach AI without overextending budgets or creating technical complexity they cannot sustain. We will explore how to start small, spend little, and scale wisely, with a focus on selecting the right problems, leveraging existing tools, and making incremental changes that compound into competitive advantage. The goal is to strip away the hype, give you an actionable path, and help you use AI to offload the inhuman work so that your business can focus on the human work that truly matters.

AI is for All Businesses, Not Just Big Tech

The misconception that AI is out of reach for SMBs is rooted in outdated assumptions. In the early days of AI adoption, the barriers were real. Developing AI models required expensive infrastructure, vast amounts of proprietary data, and highly specialised engineering talent. This meant only large corporations with deep pockets could play.

However, the democratisation of AI has completely changed this equation. Today:

- Cloud providers offer AI services on pay-as-you-go models.

- No-code and low-code AI platforms remove the need for specialised programming.
- Many AI tools come with free tiers or low-cost subscriptions.
- Public datasets and pre-trained models eliminate the need for starting from scratch.

The result is that SMBs can now deploy AI-powered capabilities in days or weeks, not years, and for hundreds of dollars, not millions. More importantly, they can focus on *applying* AI rather than *building* AI.

For example, Temple & Webster, a listed Australian online furniture retailer, embraced AI across functions ranging from product descriptions to customer support and shipping cost calculations. By 2025, AI was handling 80% of customer support, resulting in a 60% reduction in care costs relative to revenue. The results? Sales increased 21% to $600 million, and net profit quintupled to $11.3 million. The CEO emphasised that AI is now essential for maintaining competitiveness.

In this new reality, the question is no longer whether SMBs can afford AI. The real question is whether they can afford not to use it. The risk is that if you wait too long, competitors will have already embedded AI into their operations, creating efficiency and customer service gaps that will be difficult to close.

The Need To Start Small, But Smart

One of the most common mistakes SMBs make is to either do nothing or attempt too much at once. On one end of the spectrum, the perceived complexity leads to paralysis. On the other, leaders get caught up in the excitement of AI and embark on large, unfocused projects that drain resources and deliver minimal value.

The smarter approach is to start small, but with purpose. The aim is not to dabble randomly but to choose starting points that are strategically aligned with your biggest pain points or growth opportunities. This way, the early wins are not just proof-of-concepts; they are measurable improvements that free up resources for the next phase.

Start with a Strategy - Not a Shiny Tool

The AI marketplace is overflowing with tools that promise miracles. But chasing tools without a clear strategy almost always leads to disappointment. Before you buy anything or sign up for any platform, you need to answer three key questions:

1. **Where are we losing the most time, money, or customer goodwill?**
   Look for repetitive tasks that consume staff hours, processes where errors are frequent, or service gaps that frustrate customers.

2. **What outcomes do we want?**
   Are you trying to reduce costs, increase revenue, improve customer satisfaction, or free up capacity for your team?

3. **Who in our team is curious and open to**

experimenting?
Identify employees who are eager to learn and adapt. These people will become your internal champions for AI adoption.

By defining needs and framing problems to solve before selecting tools, you avoid buying technology that looks impressive but fails to address the real problem. The focus should be on value creation, not on how many AI features you can stack into your business.

Use What You Already Have - or What's Free

Many SMBs overlook the fact that they may already own software with built-in AI capabilities. CRM systems, accounting platforms, marketing tools, and productivity suites now come bundled with AI features that can automate workflows, provide insights, and even generate content.

Examples:

- Microsoft 365 and Google Workspace have embedded AI assistants.
- HubSpot and Salesforce offer AI-based sales forecasting and email optimisation.
- Xero and QuickBooks include automated reconciliation and invoice data extraction.

Before investing in new software, audit your current tools and explore any AI modules or integrations they offer.

When new tools are required, start with free or

freemium options. Many AI solutions offer robust capabilities at no cost for smaller teams or limited usage. For example:

- **ChatGPT** (free tier for basic queries and ideation).
- **Zapier** and **Make.com** for workflow automation.
- **Canva** for AI-assisted design and content generation.

The goal is to experiment without committing to large contracts. Once you see measurable benefits, you can decide whether a paid upgrade is warranted.

Seek the Right Help - Without Hiring a CTO

Hiring a full-time Chief Technology Officer is often unrealistic for SMBs. The good news is that you do not need one to start your AI journey. You can tap into expertise in flexible, cost-effective ways:

- **Freelance platforms** like Upwork and Toptal to find AI developers, data analysts, and automation specialists for short-term projects.
- **Universities and vocational institutions** to access students and interns who can work on AI projects in exchange for academic credit.
- **Local business chambers or councils** to discover grants and programs supporting digital adoption.
- **AI concierge tools** like Bardeen, Zapier,

and Make.com to automate processes without custom coding.

- **Trusted independent advisors** you can find via LinkedIn, targeting consultants with SMB and AI experience (e.g., AIQUA Solutions).

By assembling support as needed, you get access to expertise without the fixed overhead of a permanent technical team.

Target Micro-Changes, Not Mega Projects

The temptation to overhaul an entire function with AI in one go is strong, but it is rarely successful for SMBs. A better approach is to focus on micro-changes that require minimal disruption yet deliver immediate results. Over time, these small wins compound into significant operational improvement.

Examples of micro-changes:

- Automatically tagging and categorising CRM contacts.
- Using OCR (Optical Character Recognition) to digitise supplier invoices.
- Auto-generating weekly performance reports from your sales data.
- Scheduling social media posts based on AI-driven engagement predictions.

Each of these takes a fraction of the time and cost of a large project, but each also frees up hours and improves accuracy.

Prioritise Functions Ripe for AI

Certain business functions consistently deliver high returns on AI investment, especially for SMBs:

- **Sales:** Lead scoring, follow-up reminders, and proposal drafting.
- **Customer Service:** Chatbots for FAQs, ticket routing, and sentiment analysis.
- **Marketing:** Automated email campaigns, ad targeting, and content creation.
- **Operations:** Inventory forecasting, supply chain optimisation, and task scheduling.
- **Finance:** Automated reconciliation, expense categorisation, and fraud detection.

Starting in these areas increases the likelihood of early wins, as the AI tools here are mature, affordable, and easy to integrate.

Strategic Imperative

AI no longer sits on a distant horizon for small and medium-sized businesses - it is here, accessible, and increasingly vital to survival. Yet, the key to unlocking its benefits is not in chasing the biggest, most complex solutions, but in adopting a steady, value-led approach. By starting with a clear strategy, leveraging what is already available, and targeting high-impact, low-cost changes, SMBs can gain a competitive edge without overextending themselves.

The reality is that AI is at its most transformative when it quietly amplifies existing strengths, removes unnecessary friction, and frees humans to focus on what they do best. For SMB leaders, the journey is less about building an AI empire and more about nurturing a set of practical, evolving capabilities that serve the business's unique goals.

In the years ahead, those who treat AI not as a luxury but as a core operational skill will navigate uncertainty with agility and confidence. The real advantage lies in starting smart, spending wisely, and scaling only when it adds genuine value - transforming AI from an intimidating concept into a practical, everyday partner in growth.

LEVERAGING WHAT YOU ALREADY HAVE

Hidden AI in Plain Sight

"You're probably already using AI-just look closer."

Artificial Intelligence is often framed as something futuristic, expensive, and out of reach for the average business. Executives and managers are frequently led to believe that embarking on an AI journey requires purchasing new platforms, hiring highly specialised data scientists, or undergoing complex digital transformations that can consume millions of dollars. This perception has created a psychological barrier, particularly among small and medium-sized businesses, but also in enterprises that feel weighed down by their existing infrastructure.

The truth is far more encouraging: most businesses are already using AI every day, often without realising it. From the spam filter that sorts your inbox, to predictive text on smartphones, to advanced features quietly embedded in Microsoft Office 365 or Google Workspace, AI is not something sitting on the distant horizon. It is here. It is working quietly in the background, and it can be harnessed to create immediate business value.

This chapter explores the concept of "hidden AI"-the intelligence already built into the tools organisations own but rarely maximise. By uncovering and strategically deploying these features, businesses can gain quick wins, save costs, and build confidence in AI adoption, without the pressure of massive new investment.

The Misconception: AI as a Big-Bang Investment

One of the most common misconceptions about AI is that adoption requires sweeping, disruptive change. This is partly fuelled by vendor marketing, where new products are branded as "AI-powered," creating the impression that unless you buy the latest platform, you are missing out on the revolution.

In reality, this misconception blinds many leaders to the value already within reach. AI does not need to begin with high-cost innovation labs or enterprise-scale transformations. It often starts with an internal audit: asking what you already own, what AI functions are already embedded, and how those functions are being used-or overlooked.

Consider a simple but powerful truth: email spam

filters are among the most widely deployed AI applications in the world. They use machine learning to analyse billions of messages, identifying patterns of malicious or irrelevant content and protecting users automatically. Most organisations do not think of this as AI adoption, yet it is precisely that.

This recognition is liberating. It reframes AI as less about distant, intimidating transformation and more about uncovering hidden efficiencies already at your fingertips.

Everyday AI in Office Productivity Tools

Modern office suites are perhaps the most underappreciated source of AI capabilities. Both Microsoft Office 365 and Google Workspace embed advanced machine learning features that many employees barely notice, yet these tools can meaningfully transform workflows.

Microsoft Office 365

- **Excel Ideas:** This feature analyses data sets and generates automated insights. For example, a sales manager can upload raw transaction data and, within seconds, receive AI-generated charts, summaries, and trend analyses-tasks that previously took hours of manual pivot tables.

- **Insert Data from Picture:** A simple photograph of a printed table can be converted into structured Excel data through computer vision. For SMBs that rely on legacy paper-based processes, this alone can save time and reduce human error.

- **PowerPoint Designer:** AI automatically suggests slide designs, graphics, and layouts based on the text you input, dramatically reducing the time spent on presentation formatting.

Google Workspace

- **Smart Compose and Smart Reply in Gmail:** Predictive typing and suggested responses save minutes per email, which, across a team, can translate into hundreds of hours per year.

- **Explore in Google Sheets:** Similar to Excel Ideas, Explore allows users to ask natural language questions like "What are my average sales by region?" and instantly get results without advanced formulas.

These features are already paid for as part of licensing agreements. Yet surveys show that fewer than 20% of employees use them consistently. The cost is not financial-it is awareness and training.

Hidden AI in Communications and Collaboration

Beyond productivity suites, communication platforms are embedding intelligence in subtle but powerful ways.

- **Email Spam Filters:** Outlook, Gmail, and Yahoo use advanced AI models to reduce noise and security risks.

- **Noise Cancellation in Video Calls:** Zoom,

Microsoft Teams, and Google Meet all deploy AI algorithms to filter background noise. What might feel like a minor convenience is actually a deep neural network at work.

- **Real-Time Translation:** Microsoft Teams and Zoom can translate live captions across multiple languages, breaking down barriers for global teams without needing expensive interpreters.

Consider the case of LanguageLoop, a small translation services provider in Melbourne. By combining Teams' AI-driven captioning with human review, they reduced their project costs by 40% and expanded into new customer segments that previously seemed inaccessible. Their success did not come from buying new AI platforms, but from fully using the tools they already licensed.

AI in Smartphones and Everyday Utilities

Most professionals carry in their pocket one of the most advanced AI devices ever created-the smartphone.

- **Predictive Text and Autocorrect:** Based on machine learning models trained on millions of inputs.

- **Voice Assistants:** Siri, Google Assistant, and Alexa perform natural language understanding tasks, answering questions, sending messages, and integrating with third-party apps.

- **Photo Management:** Apple Photos and Google Photos use AI for facial recognition, object tagging, and smart search.

For SMBs, this is particularly powerful. A café owner, for example, can use Google Lens (a feature embedded in most Android phones) to scan invoices, identify products, and compare supplier prices instantly. What once required manual tracking can now be automated with tools already in hand.

Starting with an Audit: The First Step

The most effective way to leverage hidden AI is through a technology audit. This process does not require consultants or heavy investment. It requires curiosity, documentation, and exploration.

Audit Framework:

1. **List all software currently in use** across the organisation.

2. **Identify AI features embedded** in each tool (using vendor documentation).

3. **Assess usage levels**-are these features being applied consistently?

4. **Measure potential gains**-time saved, error reduction, improved decision-making.

5. **Prioritise quick wins** that can be rolled out with minimal disruption.

One Australian firm, Grant Day James, discovered

during such an audit that their existing Xero accounting software already had built-in AI features for expense categorisation and bank reconciliation. By training staff to use these features more effectively, they reduced month-end closing time by 30%, without adding a single new software subscription.

Case Studies Of Leveraging Hidden Ai

Case Study 1: Retail SMB Unlocking Excel AI

A family-run clothing retailer in Sydney had years of customer data but no capacity for deep analytics. By training staff to use Excel Ideas and Power BI (already included in their Microsoft subscription), they uncovered insights about buying patterns by season and age group. This led to optimised stock ordering and a 15% reduction in excess inventory within the first year.

Case Study 2: Legal Firm and Document Automation

A boutique law practice in Brisbane discovered that Microsoft Word's built-in Dictate and Editor functions could reduce drafting time by 20%. Combined with template automation, they increased lawyer productivity without hiring paralegals.

Case Study 3: Hospitality and Noise-Cancelling AI

A hotel chain in Victoria adopted AI-enabled call centre software (already part of their cloud PBX system). Noise cancellation reduced customer frustration, and sentiment analysis flagged unhappy guests in real time. Customer satisfaction scores increased by 18% within six months.

The Opportunity: Leveraging Without Spending

The greatest opportunity of hidden AI lies in its ability to provide organisations with AI-enabled transformation at zero net new cost. By embedding this mindset into leadership discussions, businesses can change the conversation from "we cannot afford AI" to "how do we maximise what we already own?"

This approach is especially valuable for SMBs but also resonates with large enterprises trying to scale adoption. Every unused AI feature represents not just lost productivity, but also a missed chance to build employee familiarity and confidence with AI.

Strategic Imperative

AI adoption does not always begin with new platforms, big investments, or grand strategies. Often, it starts with awareness and discipline. By recognising the hidden AI already embedded in everyday tools, businesses can unlock immediate value, build a culture of confidence, and create a foundation for scaling more advanced initiatives in the future.

The next wave of AI adoption will not just belong to those who build cutting-edge models from scratch. It will also belong to those who look closer, leverage what they already have, and make the invisible visible.

BRIDGING THE SKILLS GAP

Train for Curiosity, Not Mastery

"AI fluency doesn't require everyone to code. But everyone should understand what's possible."

When businesses discuss AI, one of the first barriers raised is skills. Leaders voice concerns about not having the technical know-how, not being able to hire elusive data scientists, or fearing the "talent war" that only big technology firms and consultancies can win. This anxiety creates inertia. Small and medium businesses believe they cannot compete for expertise. Larger enterprises believe only specialist teams can drive AI projects. Both perspectives miss the point.

The real skills gap is not coding. It is curiosity.

Organisations do not need every employee to master machine learning algorithms. They need employees who understand what AI can and cannot do, who can spot opportunities in their daily work, and who are motivated to experiment. In short, AI fluency matters more than AI mastery.

This chapter reframes the narrative around skills and offers a practical roadmap for businesses of all sizes to overcome this perceived deficit. By cultivating curiosity, building accessible training pathways, and leveraging external partnerships, organisations can democratise AI knowledge, unlocking transformation without waiting for a new generation of experts to arrive.

The Myth Of The Ai Skills Shortage

The conversation about AI talent often begins with dire statistics. Reports highlight the scarcity of machine learning engineers, the shortage of AI researchers, and the escalating salaries commanded by top graduates. While these numbers are true at the frontier of AI research, they do not define the reality for most organisations.

For businesses outside of Silicon Valley or Shenzhen, the pressing need is not PhD-level expertise. It is capability in applying off-the-shelf tools, integrating AI features already embedded in existing software, and rethinking workflows to include automation. The skills deficit is overstated because it frames the challenge as recruiting elite talent. The real opportunity lies in re-skilling existing teams.

Consider Shopify. When AI-powered commerce tools became mainstream, Shopify did not attempt to out-hire Google's AI research team. Instead, they invested in making their platform accessible to non-technical entrepreneurs. The result was not just product adoption, but a customer base educated in AI-enhanced commerce without advanced technical training.

The lesson is clear: mastery is not the barrier. Fluency is.

Ai Fluency Vs. Ai Mastery

AI Mastery is about writing algorithms, building models, and advancing research. It belongs to the realm of data scientists, machine learning engineers, and academics.
AI Fluency is about understanding what tools exist, where they can be applied, and how they can deliver value in context.

AI fluency allows a marketing manager to recognise when a generative AI tool can create campaign drafts. It enables a finance analyst to understand how anomaly detection can flag suspicious transactions. It equips a sales team to appreciate the role of predictive analytics in lead scoring.

The distinction matters. Without fluency, employees dismiss AI as irrelevant or too complex. With fluency, they become active participants in identifying opportunities for automation and innovation.

A simple parallel can be drawn with spreadsheets.

Most business users cannot code the underlying mathematics of Excel. Yet millions leverage Excel every day because they understand its possibilities. The same standard should apply to AI. Employees should not need to design models. They should simply understand what can be achieved.

Busting The Skills Gap Barrier

Bridging the skills gap requires more than platitudes. It requires structured interventions across four fronts: training, partnerships, culture, and champions.

Training the Workforce You Already Have

The most accessible and impactful strategy is to train existing employees. Platforms such as Google's *AI Essentials*, LinkedIn Learning's *AI for Business*, and Coursera's AI literacy programs provide structured learning without requiring deep technical backgrounds.

For example:

- **Google AI Essentials** provides 5–10 hours of content introducing core AI concepts and practical applications in business.

- **LinkedIn Learning** courses on prompt engineering or AI productivity tools are tailored to professionals.

- **Coursera and edX** offer industry partnerships, such as Andrew Ng's *AI for Everyone*, which is designed to be non-technical yet practical.

At Telstra in Australia, employees are encouraged to undertake short-form digital training courses to build AI literacy. The goal is not mastery, but equipping staff to ask the right questions and embrace experimentation.

Leveraging External Expertise

When internal training is insufficient, external experts can provide targeted support. Consultants and industry associations often run short workshops or immersion sessions. In Australia, organisations such as the AI Collective and AIIA host industry briefings, while independent consultancies deliver AI enablement programs for business leaders.

The advantage of external expertise is fresh perspective. Teams often cannot see opportunities within their own workflows. A consultant who has seen AI applied across industries can highlight immediate wins, from automating document processing to using generative AI in customer engagement.

Collaborating with Universities

Universities remain an under-utilised partner. Many have applied research centres eager for real-world problems to test. SMBs in particular can access student interns who bring both enthusiasm and fresh technical skills.

Consider Swinburne University's partnerships with Victorian manufacturing firms. Students work alongside factory staff to identify automation opportunities, often introducing AI-driven analytics tools that improve efficiency. The firms benefit from innovation without heavy recruitment costs. The students gain industry-ready experience.

Government Initiatives

Governments recognise the AI skills challenge and are investing in closing the gap. The Australian Government's **AI Adopt Program** provides free training modules and grants for SMBs exploring AI. The European Union has similar initiatives under its *Digital Europe Programme*. In the United States, the National AI Initiative Act has funded AI literacy programs for small business owners.

Businesses should view government programs not as bureaucratic hurdles but as accelerators. Free training, co-funded pilots, and subsidised workshops can reduce both cost and risk in AI adoption.

Building A Culture Of Knowledge Sharing

Skills are not just acquired through formal training. They are spread through culture. Organisations that build communities of practice accelerate learning.

Internal AI Working Groups

Creating voluntary AI working groups encourages employees to share discoveries, test tools, and report outcomes. For example, at Woolworths, "digital guilds" were established where staff across departments discuss new technologies, including AI. Such forums demystify the topic and foster a sense of shared ownership.

Digital Collaboration Tools

Platforms like Slack or Microsoft Teams can become powerful vehicles for knowledge-sharing if used deliberately. Creating AI-focused channels allows employees to exchange use cases, post articles, and share experiments. Over time, this builds collective literacy.

Learning in the Flow of Work

The most effective training is contextual. Embedding AI micro-tutorials within existing tools ensures employees learn while working. Microsoft Viva, for instance, integrates training prompts directly into workflow, reducing the friction of leaving tasks to attend workshops.

The Role of the AI Champion

Perhaps the most powerful driver of fluency is appointing an AI Champion. This individual is not necessarily the most technical person. Instead, they are cross-functional, trusted, and digitally curious.

The AI Champion acts as translator between business and technology. They encourage experiments, gather case studies, and advocate for adoption. Importantly, they humanise AI, showing colleagues that it is not an abstract force but a practical tool.

At Qantas, a senior operations manager was appointed as AI Champion during trials of predictive maintenance systems. Their credibility with frontline staff built trust in the technology, accelerating adoption and reducing resistance.

The AI Champion concept scales. In SMBs, the

founder or a digitally minded employee often plays this role. In large enterprises, each business unit may designate a champion to tailor initiatives to context.

Framework For Bridging The Ai Skills Gap

Organisations can adopt a three-tier framework:

1. **Curiosity** – Spark interest and reduce fear through exposure to success stories and accessible tools.
2. **Fluency** – Provide structured, non-technical training that enables employees to identify opportunities.
3. **Expertise on Demand** – Bring in external specialists or interns when technical depth is needed, without over-hiring.

This layered approach ensures widespread literacy while maintaining access to deeper capabilities when required.

Skills Diffusion Model

A simple way to visualise the spread of skills is the following:

- At the centre, the AI Champion represents leadership.
- Surrounding this, the Curiosity and Fluency Ring represents trained employees sharing

knowledge.

- Finally, the Outer Ring of External Expertise represents consultants, universities, and government programs.

This concentric diffusion shows how AI understanding radiates outward, reducing dependence on scarce experts.

Case Studies

Case Study 1: Commonwealth Bank of Australia (CBA)

CBA did not attempt to make every employee a data scientist. Instead, they invested in digital academies offering training modules in AI literacy, cybersecurity, and automation. Over 40,000 employees completed AI-related courses, creating a workforce fluent in identifying opportunities. The result: faster adoption of customer-facing chatbots and fraud detection systems.

Case Study 2: Bendigo and Adelaide Bank

A mid-sized Australian bank, Bendigo faced the challenge of competing with larger banks in digital services. By partnering with La Trobe University for AI projects and investing in staff training programs, they embedded AI-driven credit assessment tools while keeping their workforce engaged.

Case Study 3: Atlassian

Atlassian empowers teams through "ShipIt Days"-hackathon-style events where employees experiment with new tools. Several AI-enabled features in Jira and Confluence originated from these grassroots experiments. The company's culture of curiosity has proven more impactful than centralised top-down directives.

Strategic Imperative

The greatest barrier to AI adoption is not cost or complexity. It is mindset. Too many organisations believe they cannot act until they hire an army of specialists. This belief paralyses innovation.

The reality is different. By focusing on curiosity, training for fluency, and leveraging external partnerships, businesses can close the skills gap from within. Employees do not need to master coding. They need to recognise possibility.

AI fluency should become as universal as spreadsheet literacy. With the right cultural foundations, every employee can contribute to the AI journey. The role of leaders is to create the environment, provide the tools, and appoint champions who spread confidence.

As AI reshapes the economy, those who democratise knowledge will outpace those who hoard expertise. The future belongs to organisations that treat AI not as a specialist pursuit, but as a shared capability.

AI'S DOUBLE-EDGE

Progress with Empathy and Equity

"While AI can enhance operational efficiency, it cannot replace the nuanced understanding and emotional intelligence that humans bring to the table."

Artificial Intelligence has become a defining force of the modern era. It powers logistics networks, personalises customer experiences, accelerates medical research, and fuels new business models. The allure of AI lies in its promise of operational efficiency and unprecedented analytical capabilities. For many executives, it represents a once-in-a-generation opportunity to transform productivity and redefine competitive advantage.

Yet, alongside its benefits, AI carries profound risks and responsibilities. The very features that make it powerful - its ability to automate, classify, and optimise - can also perpetuate discrimination, diminish human connection, and create unintended inequities. AI exposes a paradox: it is both a tool of progress and a mirror of society's biases.

This chapter explores the double-edged nature of AI, asking: How can leaders leverage automation and intelligence without sacrificing empathy, fairness, and equity? How do organisations balance technological efficiency with human understanding? The answers lie not in rejecting AI but in integrating it responsibly, with empathy at the core.

The Promise of Efficiency - and Its Limits

Executives often approach AI with a singular focus: productivity. AI systems can review thousands of documents in minutes, predict consumer behaviour with precision, and identify patterns invisible to the human eye. For SMBs, these efficiencies are transformative - levelling the playing field against larger incumbents with deep resources.

But efficiency is not the whole story. Many customer interactions, employee relationships, and strategic decisions hinge not only on speed and accuracy, but also on judgement, empathy, and trust. AI can help diagnose a medical condition faster, but delivering the diagnosis compassionately remains a human responsibility. A chatbot can triage customer complaints at scale, but it cannot recognise when frustration turns into despair.

Efficiency is critical -but insufficient. Leaders must expand their success metrics beyond cost savings to include fairness, trust, and human well-being.

Bias and Discrimination: The Risk of Encoding Inequity

AI systems are only as good as the data they are trained on. Historical data reflects historical biases - ageism, racism, sexism, income-based discrimination - that, if unchecked, AI will not just reflect but amplify.

- The Workday Case (2024): Workday Inc., a major provider of HR software, faced a lawsuit alleging that its AI-driven hiring tools systematically discriminated against older applicants and racial minorities. Though Workday denied wrongdoing, the case highlighted the risk of algorithmic gatekeeping in recruitment - where biases hidden in data could deny equal opportunity at scale.

- The Mary Louis Case (2022): In another legal challenge, Mary Louis, an African American woman , alleged she was unfairly screened out by an algorithm violating housing discrimination laws. The suit argued that the automation of bias made discrimination harder to contest, as decisions were cloaked in algorithmic "objectivity."

These examples demonstrate that bias did not originate with AI - it has long existed in human decision-making. However, AI's reach and automation magnify the impact. A biased recruiter may affect dozens of candidates; a biased algorithm could exclude thousands in minutes.

For SMBs, the lesson is clear: adopting AI

without vigilance risks importing inequity into core processes. Responsible design, diverse training datasets, and regular audits are not optional-they are ethical and business imperatives.

When Empathy Goes Missing: The Klarna Experience

The Swedish fintech Klarna experimented with replacing much of its customer service with AI-driven chatbots. Initially, the system promised faster response times and reduced operational costs. Yet, by 2023, customer complaints began to rise. Clients reported frustration at being "trapped" in endless AI loops, unable to escalate issues or receive empathetic human responses.

Eventually, Klarna had to reintroduce human agents for sensitive and complex cases, acknowledging that empathy is not an optional feature but a core driver of customer loyalty.

This experience echoes a broader truth: AI struggles with emotional nuance. It can mimic patterns of empathy, but it cannot genuinely perceive or feel. For businesses, especially SMBs reliant on reputation and trust, the cost of losing empathy can outweigh the savings from automation.

Here in Australia, CommonWealth Bank recently reversed its decision to rely entirely on AI led automation in its customer service channels. It revised a retrenchment plan in favour of human intervention in customer service.

Jobs: Transformation, Not Just Displacement

Fears about job loss dominate AI debates. Indeed, history shows that technological disruption renders some roles obsolete:

- Encyclopedia salespeople disappeared with Wikipedia.
- Stenographers declined with digital transcription.
- Bank tellers diminished as ATMs proliferated and ATMs dwindled as cash less transactions became popular.

Today, AI is reshaping roles such as:

- **Receptionists:** Automated scheduling and chatbots.
- **Paralegals:** Contract analysis and legal research.
- **Programmers:** AI-assisted code generation.
- **Radiologists:** AI image analysis.

Yet, history also shows that technology creates new jobs:

- **Telehealth coordinators** emerged during the pandemic.
- **Drone operators** now support agriculture and logistics.
- **Digital asset creators** design for virtual worlds.

The key lies in redefinition, not replacement. Jobs

evolve, skills shift, and new categories emerge. SMBs can benefit by reskilling employees proactively, turning disruption into opportunity.

The Imperative of Continuous Learning

In an AI-augmented world, skills depreciate faster. The half-life of knowledge - once measured in decades - is now closer to five years or less in digital domains. For SMBs and large enterprises alike, the solution lies in continuous learning, upskilling, and cross-skilling.

- **Upskilling:** Training employees in new digital tools.
- **Cross-skilling:** Equipping staff with adjacent capabilities (e.g., a marketing executive learning basic data analysis).
- **Learning Mindset:** Encouraging curiosity over mastery.

This aligns with the previous argument: fluency matters more than code. Employees need not become AI engineers but must understand what's possible-and how to collaborate effectively with machines.

Striking the Balance: Technology and Humanity

To navigate AI's double-edge, leaders must embrace a balanced approach:

1. **Leverage AI's Strengths:** Data processing, efficiency, pattern recognition.

2. **Preserve Human Qualities:** Empathy, ethical reasoning, nuanced understanding.

3. **Embed Guardrails:** Audit algorithms for bias, build transparent processes, and ensure accountability.

4. **Redefine Metrics of Success:** From cost savings to fairness, trust, and customer experience.

SMBs, with their flatter structures and closer customer relationships, are uniquely positioned to strike this balance more nimbly than large corporates.

Strategic Imperative

AI is not destiny - it is a design choice. The way organisations integrate automation will shape whether it becomes a force of progress or a vehicle for inequity.

The most effective leaders will be those who marry efficiency with empathy, embedding fairness and emotional intelligence at the core of digital transformation. AI may optimise processes, but it is humans who define values, interpret meaning, and deliver compassion.

In the final analysis, the true test of AI adoption will not be how much it saves or how fast it scales, but whether it uplifts human potential without erasing humanity's essence.

GUARDING YOUR DATA

Privacy, Security, and Trust in the AI Era

"AI is powerful - but it's only as safe as your inputs and policies."

Artificial Intelligence is no longer at the periphery of business operations; it is deeply embedded in decision-making, customer interactions, product design, logistics, and financial analysis. With this integration, however, comes an exponential rise in both the volume and sensitivity of data being generated, processed, and shared. Data has always been a valuable corporate asset, but in the AI era it has become the raw fuel that powers predictive engines, recommendation systems, and generative tools.

Yet this power is double-edged. The same systems that enable unprecedented productivity and insights also expose organisations to new vulnerabilities: data leakage, intellectual property

(IP) theft, regulatory breaches, and the erosion of stakeholder trust. One misstep can result not only in financial losses but also in reputational damage that may take years to repair.

The central lesson is simple: AI is only as safe as the data you feed it and the governance structures you put in place. This chapter explores the risks, safeguards, and strategic actions needed to protect privacy, security, and trust in the AI-driven enterprise.

Why Ai Raises The Stakes

Unlike traditional software, AI systems learn from (and sometimes retain) inputs. This means data doesn't just "flow through" a system; it can shape models, outputs, and even future applications. The risks are not hypothetical. In the last two years, several companies have been embarrassed by employees unintentionally leaking customer data or trade secrets into public AI systems.

Three critical risks stand out:

1. **Data Leakage:** Sensitive information can be inadvertently exposed if entered into public AI models. Once submitted, it may be stored or even used for further training.

2. **Loss of Intellectual Property (IP):** Proprietary data or product designs used in prompts could be captured by third-party systems. Competitors may indirectly benefit if safeguards are not in place.

3. **Hallucinations:** AI-generated "confident but wrong" outputs can mislead decisions, damage customer trust, or misrepresent facts, especially in regulated industries like finance, healthcare, or legal services.

Case Example: Samsung's Code Leak

In 2023, engineers at Samsung uploaded confidential source code into ChatGPT to debug issues. That code - part of Samsung's proprietary semiconductor designs - was at risk of being absorbed into the AI model. While OpenAI later implemented clearer policies, the incident became a cautionary tale: well-intentioned use can create high-stakes vulnerabilities.

The Pillars Of Responsible Ai Data Use

Data Privacy

AI systems must align with regional and international data protection laws. In the EU, the General Data Protection Regulation (GDPR) enforces strict requirements for consent, processing, and storage. In Australia, the Privacy Act is undergoing reforms to strengthen accountability for AI-driven data use. In the US, multiple state-level frameworks (California's CCPA being the most notable) are converging toward similar principles.

Security

Traditional cybersecurity remains essential, but AI introduces new attack surfaces: model inversion (extracting training data), prompt injection attacks, and adversarial inputs designed to trick models into unsafe outputs. These risks demand encryption, access controls, monitoring, and frequent audits.

Trust

Trust is not just about compliance; it is about stakeholder confidence. Customers must feel their personal data is safe. Employees must trust that their inputs into AI tools will not be weaponised against them. Shareholders must see evidence that the company has governance frameworks robust enough to withstand scrutiny.

Practical Safeguards For Smaller Businesses

1. Start with Trusted Vendors

Not all AI tools are created equal. Before adopting, organisations must ask vendors:

- How is data collected, stored, and used?
- Are prompts logged? If so, where and for how long?
- Are inputs used to further train the model?
- What compliance certifications do they hold (ISO 27001, SOC 2, GDPR, HIPAA)?

2. Develop a Clear AI Use Policy

A company-wide AI Use Policy should outline:

- **Permitted Tools:** What AI applications are

approved.

- **Data Restrictions:** Explicit lists of data that must never be shared (e.g., customer lists, health records, financial forecasts, source code).

- **Accountability:** Who reviews and approves outputs in sensitive workflows.

- **Review Mechanism:** How often the policy is updated as tools evolve.

This transforms AI adoption from ad hoc experimentation into a governed, risk-managed practice.

3. Human-in-the-Loop Workflows

AI should be an assistant, not an unchecked authority. By embedding humans in the approval process - particularly for high-stakes outputs such as financial reports, legal documents, or medical recommendations - organisations reduce the likelihood of harm from hallucinations or misinterpretations.

4. Enterprise-Grade Tools

For sensitive environments, consumer-grade AI tools are insufficient. Alternatives include:

- **ChatGPT Enterprise / Team**: Does not train on customer prompts.

- **Microsoft Copilot for 365**: Enterprise-grade compliance aligned with Microsoft's security frameworks.

- **Local Model Deployments:** Tools like LLaMA or open-source models that can be hosted within a company's private environment.

5. Encryption and Regular Security Audits

Every input, output, and dataset should be encrypted in storage and in transit. In addition, routine penetration tests and red-team exercises should be conducted to simulate adversarial AI attacks.

Sector-Specific Risks And Solutions

Financial Services

- **Risk:** AI systems recommending products may unintentionally discriminate (e.g., offering worse terms to certain demographics).

- **Safeguard:** Bias testing, explainability protocols, and regulatory audits.

Healthcare

- **Risk:** Patient records exposed in training data could violate HIPAA or Australian privacy laws.

- **Safeguard:** Use de-identified data, enterprise tools, and strict consent frameworks.

Retail and E-commerce

- **Risk:** Over-reliance on chatbots may frustrate

customers if hallucinations occur.

- **Safeguard:** AI-human handoff processes and customer feedback monitoring.

Building A Culture Of Data Responsibility

Training Employees

Employees must understand:

- The risks of pasting sensitive data into public tools.
- The basics of prompt engineering and its impact on outputs.
- The company's AI use policy.

Encouraging Data Literacy

Workshops and internal training should go beyond technical rules to build data awareness: how privacy works, what compliance requires, and why trust matters.

Appointing Data Stewards

Much like "AI Champions" described earlier, appointing Data Stewards ensures ongoing vigilance. Their role: monitor compliance, guide employees, and liaise with security teams.

The Trust Dividend

Organisations that safeguard data do more than avoid risk- they create competitive advantage. Customers increasingly choose providers based on trustworthiness. By demonstrating transparent practices, clear governance, and ethical AI adoption, businesses can deepen loyalty and open new markets.

Case Example: Apple
Apple's stance on privacy - "What happens on your iPhone, stays on your iPhone" - has become a marketing differentiator. While critics argue the claim is not absolute, Apple has nonetheless positioned trust as a brand asset, winning consumer confidence even in highly competitive markets.

Looking Ahead: Regulatory Convergence and Global Standards

Governments are moving quickly to regulate AI. The EU AI Act, Australia's AI Ethics Framework, and the US NIST AI Risk Management Framework are converging toward common principles: transparency, accountability, security, and fairness. Early adopters who align with these frameworks will avoid costly retrofits and build reputational resilience. Businesses cannot rely on government to protect them, governments are known to be reactive in the past with emerging technologies. AI is everybodys responsibility as Cybersecurity is, the responsibility to safe gaurd sensitive data is everyone's responsibility, only that AI accentuates the risks if your data falls into the wrong hands.

Strategic Imperative

AI offers extraordinary opportunities - but it also multiplies risks if organisations fail to handle data responsibly. The guiding principle is simple: AI is only as safe as your inputs, policies and individual ownership.

Guarding data in the AI era is not a compliance box-ticking exercise. It is the foundation of stakeholder trust, customer loyalty, and long-term competitiveness. Organisations that protect privacy, secure their data pipelines, and embed human oversight will not just avoid harm; they will build a reputation for responsible innovation-one of the most valuable currencies in the digital economy.

TRANSFORMATION REIMAGINED

The Boardroom's Role in the AI Age

"With pervasiveness of digital technology in society, there is a skill and experience gap at leadership levels that is stifling transformation initiatives within businesses."

Digital Transformation (DT) is no longer a choice; it is a strategic imperative. The acceleration of AI, cloud adoption, data analytics, and platform ecosystems has made it impossible for organisations to survive - let alone thrive - without embedding digital at their very core. Yet despite billions in investment and untold hours of effort, two out of three digital transformations fail.

Why? Not because of technology. Not because of talent shortages alone. The primary causes are leadership blind spots, weak governance, and lack of board-level ownership. The high stakes previously mentioned of 88% and 95% failure rates for AI pilots not scaling not returning value can be directly attributed as DT failures. AI implementations will change KPIs, will introduce new governance, will shift organisation designs and force business process design. If these are not evident, the expectations from AI were not supported by the changes they demand.

Too often, DT is relegated to "just another IT project" - implemented in silos, delegated down the organisational chain, and measured only in terms of budget spend rather than business outcomes. In reality, DT is about business model reinvention and enterprise-wide cultural change, which demands strategic oversight at the very top of the organisation.

This chapter reframes the role of boards and executive leaders in the AI-driven era of transformation. It sets out three essential tests for evaluating the maturity and direction of your digital initiatives and highlights how boards must evolve - from passive overseers to active stewards of digital futures.

Why Two Out Of Three Digital Transformations Fail

The high failure rate of digital initiatives is not due to lack of ambition or resources. The primary culprits are:

- **Leadership Gaps** – Boards and senior executives often lack digital fluency, making them ill-equipped to ask the right questions or interpret risks and opportunities.

- **Change Management Failures** – Transformation disrupts entrenched cultures, processes, and KPIs. Without structured change leadership, resistance overwhelms progress.

- **Governance Weaknesses** – Many DT initiatives are launched without clear accountabilities, measurable outcomes, or alignment with business strategy.

The result is predictable: projects balloon in scope, business value remains elusive, and executives quietly reposition failure as "learning experiments".

In the AI age, this is unacceptable. Organisations cannot afford for transformation to fail at such high rates - the competitive environment is too unforgiving.

Digital Transformation As Business Model Innovation

Boards that see DT as an "IT project" will inevitably fail. Those that view it as business model innovation will succeed. History shows us that.

- **Blockbuster vs. Netflix**: Blockbuster optimised store operations; Netflix reimagined distribution and customer experience.

- **Microsoft Office vs. Office 365**: Selling software in a box was sustainable for a decade;

subscription-based cloud productivity services unlocked exponential, recurring revenue. Microsoft planned the obsolescence of its own products to re-invent itself.

- **Retail Banking**: Banks that digitised only internal workflows lagged behind those that reimagined entire customer journeys via apps, digital wallets, and embedded finance. Overnight Fin-Techs were luring their customers away.

The common thread is clear: DT succeeds when it redefines value creation and delivery, not when it simply digitises existing processes.

The Three Essential Tests For Board Oversight Of Digital Transformation

To break free from failure cycles, corporate boards and executives must subject their digital programs to three rigorous tests.

Test One: The Degree of Change Involved

A genuine digital transformation involves deep, structural change:

- Organisational design shifts.
- New governance mechanisms.
- Reimagined KPIs that measure digital engagement, customer lifetime value, or ecosystem growth - not just quarterly revenue.

- Fundamental process redesigns, often requiring automation, AI augmentation, and ecosystem integration.

- DT either fundamentally changes your outcome or changes the process that produces the outcome

Boards must ask:

- *Is this program truly transformative, or just incremental IT modernisation?*

- *Does leadership own the change, or is it buried in middle management?*

- *Are KPIs and governance mechanisms aligned to digital realities?*

When change is superficial, results will be too.

Test Two: Is the Program Interdisciplinary?

Digital is not the responsibility of IT alone. Successful DT requires:

- **Cross-functional collaboration** - technology, marketing, finance, HR, and operations working as one.

- **Creative processes** - blending design thinking, data-driven insights, and agile delivery.

- **Cultural transformation** - flattening silos, encouraging experimentation, and rewarding collaboration.

Boards must demand evidence that rigidity is being dismantled. Ask:

- *Does the DT initiative cross silos?*
- *Are diverse voices - technical, creative, operational - represented?*
- *Is executive leadership driving collaboration, or allowing entrenched barriers to persist?*

Without interdisciplinary execution, DT collapses into fragmented, tactical projects.

Test Three: Who is Driving the Transformation?

Perhaps the most critical test: who owns the transformation?

- If IT drives it, outcomes are framed in systems language, not business value.
- If consultants drive it, outcomes align with methodologies, not enterprise renewal.
- If business leadership drives it, supported by empowered digital and product leaders, transformation aligns with vision, customers, and value creation.

Boards must probe:

- *Does the CEO actively champion digital I initiatives?*
- *Are business leaders - not just IT leaders -*

accountable for results?

- *Are prioritisation decisions ruthless, ensuring resources flow to the most strategic digital bets?*

Leadership ownership is the dividing line between symbolic transformation and true reinvention.

The Ai Dimension: Raising The Stakes For Boards

AI magnifies both the opportunities and the risks of DT. Boards must now contend with:

- **Exponential efficiency gains** – AI-enabled automation, predictive analytics, and customer personalisation.

- **Ethical dilemmas** – bias, fairness, explainability, and workforce disruption.

- **New competitive landscapes** – markets redefined overnight by AI-first challengers.

Boards that fail to integrate AI fluency into their governance risk irrelevance. Fiduciary duty now includes digital and AI stewardship.

Governance Practices For The Ai Age

Boards must evolve their governance playbooks to include:

1. **Digital & AI Literacy** – Every board

member need not be a technologist, but all must be conversant enough to challenge assumptions and scrutinise risks.

2. **Strategic Alignment** – Digital and AI programs must tie directly to long-term business strategy, not sit as side projects.

3. **Risk & Ethics Oversight** – Independent review of AI risks, fairness audits, and ethical guardrails.

4. **Talent & Leadership Pipelines** – Boards must ensure executive teams include leaders capable of driving AI-enabled transformation.

5. **Outcome-Based Reporting** – Regular updates to the board should focus on business outcomes (new revenue streams, customer engagement, operational resilience), not IT metrics.

A Call to Action for Boards

The evidence is clear: boards cannot outsource digital/AI transformation to management alone. Fiduciary responsibility now extends to ensuring organisations are digitally resilient, AI-ready, and strategically agile.

Boards must shift from being late-stage reviewers to early-stage stewards of transformation. They must ask harder questions, demand sharper priorities, and embed digital thinking into every dimension of governance.

In the AI age, the success of a transformation program will increasingly be determined not by technical capacity, but by boardroom courage, digital literacy, and leadership alignment.

Digital transformation in the AI era is not about adding new applications or upgrading infrastructure. It is about reimagining the very DNA of the enterprise- how value is created, delivered, sustained and governed. Despite years of effort, two out of three transformations still fail, most often because boards and executives abdicate their responsibility to lead, govern, and steward change. Dangerously many organisations are centralising AI as a CAIO responsibility.

To succeed, boards must treat digital transformation as business model innovation, not just technology modernisation. Every initiative should be tested against three fundamental questions: does it create new value, does it deliver measurable outcomes, and can it be sustained at scale? Equally important, governance practices must evolve to reflect the responsibilities of the AI era, from data stewardship and ethical oversight to ensuring resilience and accountability.

Boards that rise to this challenge will convert the AI age into an era of enduring enterprise value. Those that do not risk sharing the fate of Kodak, Blockbuster, and Nokia - remembered not for their achievements but for their failure to act when it mattered most.

Strategic Imperative

This final imperative is not just about Board responsibilities and accountabilities alone it is a mirror reflecting the journey we have taken throughout this book.

We began by calling to attention the dangers of complacency and demystifying AI, stripping away the hype to reveal practical pathways for businesses. We examined hidden AI already embedded in our tools, learned how to scale responsibly, built guardrails for privacy and security, and redefined the role of leadership and boards in steering transformation.

Now, standing at the threshold of the future, the lesson is simple but profound: AI is not about tomorrow-it is about today. The revolution is not loud, but it is relentless. It rewards the vigilant, the adaptable, and the courageous, while exposing the complacent.

The closing call to action for leaders is therefore not to predict the future, but to prepare for it. Not to resist, but to reimagine. Not to fear displacement, but to embrace augmentation.

The AI age will not wait. It is here, pervasive, and unstoppable. The question is no longer whether you will join it, but whether you will shape it. I wish you well.

BOOKS BY THIS AUTHOR

India's Innovation Blueprint: How The World's Largest Democracy Is Becoming An Innovation Superpower

India is the fourth largest economy by gross domestic product yet India's inequities are visible; its young population tread a thin line between opportunity and pitfall; its infrastructure has gaping holes; and it's a slow chaotic democracy. This book establishes that India is not just leveraging innovation for global competitiveness alone, but is also leveraging innovation as the specific instrument for inclusive growth. This book identifies gaps in the current innovation ecosystem and recommends a portfolio approach and calls for a National Innovation System (NIS) as a blueprint to fix the gaps. It suggests that for India to succeed in identifying, funding and sustaining a balanced innovation portfolio, India will also have to succeed in eliminating poverty, increase its rural GDP manifold, and provide employment, education and health for all its citizens.

PRAISE FOR AUTHOR

Leveraging a vast wealth of knowledge and experience, George presents a highly pragmatic approach to AI. This books cuts through the perceived complexities of AI, busting the myths surrounding it whist championing the humanity of those that it will serve. As you read, you'll be energised and empowered to take practical steps in your AI journey.

- JOSHUA GROENESTYN, DIRECTOR, SYDNEY PACKAGING

I found the pre-launch session by George to be extremely insightful. Great clarity, every point made perfect sense, especially on how we need to use it as a tool and at the same time be aware of the loopholes and the strategies to safeguard our data without compromising our work. A tool is only as good as the hands of a skilled craftsman.

- ELEANOR SYBIL D'CRUZ

Made in the USA
Monee, IL
04 November 2026

67384326R00085